T0281462

Springer Transactions in Civil and Environmental Engineering

Springer Transactions in Civil and Environmental Engineering (STICEE) publishes the latest developments in Civil and Environmental Engineering. The intent is to cover all the main branches of Civil and Environmental Engineering, both theoretical and applied, including, but not limited to: Structural Mechanics Steel Structures Concrete Structures Reinforced Cement Concrete Civil Engineering Materials Soil Mechanics Ground Improvement Geotechnical Engineering Foundation Engineering Earthquake Engineering Structural Health and Monitoring Water Resources Engineering Hydrology Solid Waste Engineering Environmental Engineering Wastewater Management Transportation Engineering Sustainable Civil Infrastructure Fluid mechanics Pavement engineering Soil dynamics Rock mechanics Timber engineering Hazardous waste disposal Instrumentation and monitoring Construction management Civil engineering construction Surveying and GIS Strength of Materials (Mechanics of Materials) Environmental geotechnics Concrete engineering timber structures Within the scopes of the series are monographs, professional books or graduate textbooks, edited volumes as well as outstanding PhD theses and books purposely devoted to support education in mechanical engineering at undergraduate and graduate levels.

More information about this series at http://www.springer.com/series/13593

Bradley Ladewig
Muayad Nadhim Zemam Al-Shaeli

Fundamentals of Membrane Bioreactors

Materials, Systems and Membrane Fouling

 Springer

Bradley Ladewig
Department of Chemical Engineering
Imperial College London
London
UK

Muayad Nadhim Zemam Al-Shaeli
Department of Chemical Engineering
Monash University
Melbourne, VIC
Australia

ISSN 2363-7633　　　　　　　ISSN 2363-7641　(electronic)
Springer Transactions in Civil and Environmental Engineering
ISBN 978-981-10-9504-7　　　ISBN 978-981-10-2014-8　(eBook)
DOI 10.1007/978-981-10-2014-8

Acknowledgements

First, I would like to thank God for blessing me with faith, strength and helping me understand myself by looking at his manifestations.

Second, it is my genuine pleasure to express my deep sense of thanks to my main supervisor Prof. Bradley Ladewig for guiding me through my writing this book. This book would never have been completed without his virtuous guidance. His timely recommendation, meticulous scrutiny, scholarly advice and scientific approach have helped me significantly to accomplish this book. I really appreciate his support, recommendation and encouragement.

I would like to acknowledge the financial support from Iraqi Government/Ministry of Higher Education and Scientific Research in Iraq for their support by granting me a Ph.D. scholarship.

I would like to extend my sincerest thanks and appreciation to my parents, my brothers and my sisters for giving me encouragement and support.

Lastly and most importantly, my deepest appreciation to my wife (Soorah) for her endless love without expecting any gifts and providing me with her constant support and recommendation by standing shoulder to shoulder with me to get though all the difficulties I faced as an international student. This work is dedicated to her and our lovely daughter (Mariam).

Contents

Chapter 1
Introduction to Membrane Bioreactors

This chapter gives a general introduction to membrane science and technology, and begins with the definition of terms and provides a description of membrane processes currently implemented in different fields. Specifically, the Membrane Bioreactor (MBR) technology is introduced and followed by a short overview of the historical development and different configuration of MBRs. Finally, the advantages and disadvantages of membrane bioreactors are discussed. To sum up, MBR acts an efficient, reliable and cost-effective technology that deals excellently with the growing demands for treating wastewater, which can then be returned to the hydrological cycle without any adverse effects.

1.1 Introduction

A membrane is defined as a thin selective barrier between two phases (gas or liquid), which is impermeable to the transfer of specific particles, molecules or substances, colloidal, and dissolved chemical species other than water or solvent (Mulder 1997). A material of reasonable mechanical strength that maintains a high throughput of a desired permeates with a high degree of selectivity is ideal for the production of membranes. Usually, a thin layer of material with a narrow range or domain of pore size and a high surface porosity affect the physical structure of the membrane. The physical structure of a thin layer membrane leads to the separation of dissolved solutes in liquid streams and the separation of gas mixtures for membrane filtration (Visvanathan et al. 2000).

Membrane processes or productions are categorised based on (1) the driving force which is used for separation of impurities such as pressure (ΔP) temperature (ΔP), concentration gradient (ΔC), partial pressure (Δp), or electrical potential (ΔE), (2) the mechanism of separation, (3) the particular application of membrane, (4) the size of the retained material, and (5) the type of membrane (Baker 2004).

© Springer Nature Singapore Pte Ltd. 2017
B. Ladewig and M.N.Z. Al-Shaeli, *Fundamentals of Membrane Bioreactors*,
Springer Transactions in Civil and Environmental Engineering,
DOI 10.1007/978-981-10-2014-8_1

Considering the categorization based on the pore size or size of retained material, membranes can be classified as ultrafiltration (UF), microfiltration (MF), nanofiltration (NF) and reverse osmosis (RO) membranes, dialysis, electrodialysis (ED), where the first four processes produce permeate and concentrate (Radjenović et al. 2008). As illustrated in Table 1.1, UF and MF are low pressure-driven processes in which feed water is driven through a synthetic micro-porous membrane and then divided into permeate (which passes through the membranes) and retentate (which includes nonpermeating species). These membrane processes are known to be more efficient in removing microorganisms and particles from wastewater. In comparison to UF and MF, RO is a high pressure-driven process used to remove dissolved constituents such as salts, low molecular organic and inorganic pollutants from waste water remaining after advanced treatment with MF. NF operates at a pressure range between RO and UF, targeting removal of divalent ion impurities (Visvanathan et al. 2000).

Table 1.1 Characteristics of membrane processes (Nath 2008; Perry and Green 1997; Koros et al. 1996)

Type of process	Size of materials retained	Driving force	Type of membrane	Application
Ultrafiltration	1–100 nm macromolecules	(ΔP) (1–10 bar)	Micro-porous	– Separation of proteins and virus – Concentration of oil in water emulsions
Microfiltration	0.1–10 μm microparticles	(ΔP) (0.5–2 bar)	Porous	Separation of bacteria and cells from solutions
Nanofiltration	0.5–5 nm molecules	(ΔP) (10–70 bar)	Micro-porous	– Separation of dye and sugar – Water softening
Reverse osmosis	<1 nm molecules	(ΔP) (10–100 bar)	Nanoporous	– Desalination of sea and brackish water – Process of water purification
Dialysis	<1 nm molecules	(ΔC)	Micro-porous or nanoporous	– Purification of blood
Electrodialysis	<1 nm molecules	(ΔE)	Micro-porous or nanoporous	Separation of electrolytes from nonelectrolyates
Pervaporation	–	(ΔC)	Nanoporous	Dehydration of ethanol and organic solvents
Gas separation	–	Partial pressure difference (1–100 bar)	Nanoporous	Hydrogen recovery from process gas streams, dehydration and separation of air
Membrane distillation	–	(ΔT)	Micro-porous	Water purification and desalination

A more recently developed membrane process is MBR, which combines MF or UF, and a bioreactor for biological treatment. This process is an emerging technology increasingly used for both municipal and industrial waste water treatment and has led to significant advances in knowledge and experience related to their design and operation (Crawford et al. 2000; Drews 2010; Hong et al. 2002; Hwang et al. 2010; Judd 2006; Judd and Judd 2010; Meng et al. 2009; Stephenson et al. 2000; Thomas et al. 2000; Visvanathan et al. 2000). It represents a good innovative process in which gravity settling or clarifiers of the conventional activated sludge system (CAS) is replaced by membrane separation module (Bella et al. 2007; Hong et al. 2002; Le-Clech et al. 2006). Such a module reproduces MF or UF processes with pore sizes ranging from 0.05 to 0.4 µm (Bouhabila et al. 2001). This enables the separation of solid–liquid and act as an advanced treatment unit for specific pollution agents (e.g. coliform bacteria or suspended solids). It must be kept in mind that these agents cannot be completely eliminated by conventional waste water treatment processes (Bella et al. 2007; Le-Clech et al. 2006)

As seen schematically in Fig. 1.1, in a conventional activated sludge (CAS) process, the microorganisms accumulate into flocs and these flocs are suspended in wastewater to facilitate treatment. Once the waste water is treated, the flocculated microorganisms must be eliminated from clean water (Hussain et al. 2010). Conventionally, a clarifier is used for liquid–solid separation; therefore, a successful treatment in CAS process relies on the development of flocs that settle well. Furthermore, conventional activated sludge process (CASP) does not cope well with fluctuations of wastewater flow rate or composition.

Conversely, in MBR, the membrane component eliminates the need for a clarifier and is performed using low-pressure MF, UF or NF membranes. Additionally, MBR systems allow the complete physical retention of bacterial flocs and almost all suspended solids (individual microorganisms, large biological flocs and inert particles) within the bioreactor and therefore can offer excellent disinfection capacity

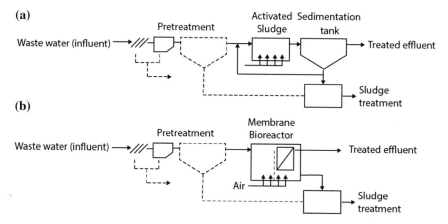

Fig. 1.1 **a** Typical process schematic for conventional activated sludge processes (CAS) and **b** membrane bioreactor processes in wastewater treatment. Adapted from Leiknes (2010)

(Le-Clech et al. 2006; Li and Chu 2003). As a result, the total coliform bacteria reduction can reach an average of log 7 (Hirani et al. 2010). Due to its advantages, MBR has been demonstrated to be highly effective for the treatment of polluted surface water supplies to produce potable water (Smith et al. 1969). Thus, in comparison to other membrane processes, MBR has the potential for the treatment of many types of wastewater; this does not solely lie in its application to biological degradation and nitrification, but also because it could replace other conventional treatment units such as flocculation, sedimentation, filtration and disinfection. In industry, MBR is used as secondary treatment in order to reduce biodegradable and non-biodegradable matter in the end product (because the presence of oxygen) or as MBR can be used as an advanced treatment to remove residual nutrients which are not fully removed during secondary treatment (Tchobanoglous et al. 2004).

1.2 History of Membrane Bioreactors

Membrane bioreactors were initially developed in the 1960s when commercial scale UF and MF membranes became available. The original process was introduced by Dorr-Olivier Inc. (Milford, Connecticut) and combined a crossflow membrane filtration loop with an activated sludge bioreactor (Enegess et al. 2003). Polymeric flat sheet (FS) membranes with pore sizes ranging from 0.003 to 0.01 μm were used in this process (Yamamoto et al. 1989). Replacing the settling tank of the conventional activated sludge process seemed to be appealing. However it was difficult to validate the use of such process due to the associated high costs of the membranes. The low economic value of the product (tertiary effluent) and potential rapid loss of performance due to fouling also impaired the process of membrane bioreactors. And as such the focus was on attaining high fluxes. Nevertheless it was important to deliver the mixed liquor suspended solids (MLSS) at a high cross flow velocity, incurring a significant energy consumption (of the order 10 kWh/m^3 product) to reduce membrane fouling.

This first generation of MBRs suffered from a substandard economic performance and as such their suitability was limited to a narrow range of applications such as ski resorts, hotels or isolated trailers parks. The major breakthrough for MBRs came in 1989 with the research conducted by Yamamoto et al. They demonstrated for the first time the use of submerged membranes in bioreactor, which was a major innovation, compared to the prior approach with the separation device located externally to the reactor. Yamamoto et al. (1989) investigated the feasibility of direct membrane separation using hollow fibre in an activated sludge aeration tank, and identified the key parameters, which give a stable operation and effective organic stabilisation and nitrogen removal. As a result, the number of membrane bioreactors used to treat municipal wastewater increased significantly. In 2005, the market value of MBRs reached $217 million and continued to increase reaching $360 million in the year 2010 (Judd 2006).

The acceptance of modest fluxes (25 % or less of those in the first generation) and the idea to use two-phase bubbly follow to limit membrane fouling has been another aspect of recent development of MBR systems. Bubbling or air scouring is used to deter clogging of the membrane modules from solid concentrations and as a technique to control membrane fouling. According to Zhang et al. (2011b), two types of bubbling can be used to control membrane fouling. The first one is a slug bubble and the second one is a free bubble. They concluded that using slug bubbles showed better antifouling performance than free bubbles in FS MBR under both short-term and long-term operation. In short-term operation, a high flux operation was achieved at 36 h with 40 L m^{-2} h^{-1}. In contrast, moderate flux operation of 14 days was possible a flux of 24 L m^{-2} h^{-1}.

From the mid-1990s there was an exponential increase in MBR plant installations as a result of lower operating cost with the submerged configuration and a continual decrease in membrane cost. Further enhancement in the designs and operation of MBRs has been achieved and these have been assimilated into larger plants. Early MBRs operated at solid retention times (SRT) as high as 100 days with MLSS up to 30 g L^{-1}. Recently, the trend is to apply a lower SRT (around 10–20 days). This leads to more manageable MLSS levels (10–15 g L^{-1}). Due to these new operating conditions, the membrane-fouling tendency in MBR has begun to decrease and overall maintenance has been simplified, as less frequent membrane cleaning is necessary (Le-Clech et al. 2006). Currently a variety of MBR systems are commercially available, mostly using submerged membranes. Although some external modules are available, these external systems also use two-bubbly phase flow for fouling mitigation.

Generally, for membrane configuration, hollow fibre and FS membrane sheets are used in the applications of MBRs (Stephenson et al. 2000). The economic feasibility of current generation MBRs depends on an attainable permeate flux, predominantly controlled by appropriate fouling control strategy with a modest energy input (usually ≤ 1 kWh/m^3 product).

Effective fouling alleviation methods can be executed only when the phenomena appearing at the membrane surface are fully understood. The plethora of publications dealing with membrane fouling and published within the last 5 years is overwhelming and can confuse many readers. To simplify the situation, a comprehensive yet concise overview of MBR foulants parameters and fouling will be presented in this book.

1.3 Membrane Bioreactor Configurations

Basically, there are two membrane configurations used in the membrane system. The first configuration is side-stream (external) membrane bioreactors (see Fig. 1.2) and the second one is submerged MBR (the membrane is immersed directly into bioreactor) (see Fig. 1.3). The second one is more applicable in wastewater treatment than the first one because it has many advantages such as lower energy

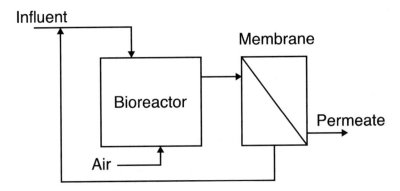

Fig. 1.2 Side-stream membrane bioreactor with external pressure-driven membrane unit. Adapted from Radjenović et al. (2008)

Fig. 1.3 Submerged Membrane Bioreactor with internal vacuum pressure-driven membrane filtration. Adapted from Radjenović et al. (2008)

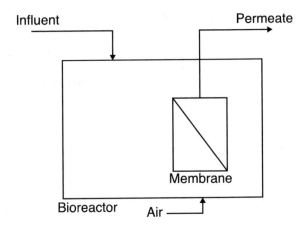

consumption, higher hydraulic efficiency and simple design. However, side-stream configuration is possibly used for wastewater treatment, with wastewater (feed) is pumped into the membrane and part of the permeate is collected while the other part is returned back to the MBR. Side-stream configuration has the capability to control membrane fouling significantly; resulting in constant flux but the energy consumption and complex design are the major limitations.

The configurations of MBR are based on geometry (either cylindrical or planar). There are five major membrane configurations currently used in practice (Radjenović et al. 2008)

1. Hollow fibre (HF)
2. Spiral-wound
3. Plate-and-frame (i.e. FS)
4. Pleated filter cartridge
5. Tubular.

The first three membrane configurations are widely used in MBRs. In the first membrane module, the large amounts of these membranes make a bundle and the ends of the fibre are sealed properly in an epoxy block connected with the outside of the housing. The water can flow from inside to outside of the membrane or vice versa. These membranes operate under both pressure and vacuum. Due to the lower energy cost and back-flushing capability, hollow fibre membranes are most popular in MBRs.

Spiral-wound membrane configuration are mostly used for RO and NF processes. The Spiral-wound membrane configuration are coiled through the perforated tube in which permeate (effluent) goes out. The standard manufacturing of the spiral-wound membrane configuration make their installation easier with less cost in membrane production. The installation of these membrane configuration can be performed in series or parallel in plants with higher capacity.

Currently, plate-and-frame membrane modules are widely used in water and wastewater treatment industry. They are composed of FS membranes with separators and/or support membranes. The pieces of these sheets are fastened onto a plate. The water flows across the membrane and the permeate is extracted through pipes which emerge from the interior of the membrane module in a process that operates under vacuum (Radjenović et al. 2008).

The last two membrane configurations are pleated filter cartridge and tubular. These membranes are not widely used in industrial scale. Obviously, tubular membranes are wrapped in a pressure vessel (tube) and then mixed liquor is pumped through them. They used specifically for side-stream MBR configurations.

1.4 Advantages and Limitations of Membrane Bioreactors

Membrane bioreactors have attracted extensive attention as a result of their numerous advantages over CASP. The advantages of MBRs include excellent treated water quality, high biodegradable efficiency, small footprint and reactor requirements, absolute biomass retention and ease of stable operation. They can also display high effluent quality, flexible operation, absolute removal of bacteria, high volumetric loading up to 20 kg COD/m^3 per day, excellent disinfection capability and turbidity less than 0.5 NTU (number transfer unit), low sludge production, compactness, enable high removal efficiency of biological oxygen demand (BOD) and chemical oxygen demand (COD) (Judd 2006, 2008; Li and Chu 2003; Liao et al. 2006; Pan et al. 2009; Wang et al. 2008; Yamato et al. 2006; Yang et al. 2006; Bouhabila et al. 2001; Chang et al. 2006; Kimura et al. 2005; Meng et al. 2009; Stephenson et al. 2000; Visvanathan et al. 2000; Zhang 2009).

As a result, the MBR process has now become a viable alternative for the treatment and reuse of municipal and industrial wastewaters. MBRs are therefore considered a promising tool for future waste water treatment (Rosenberger and Kraume 2002; Rosenberger et al. 2002; Chu and Li 2005; Côté et al. 1998; Davis et al. 1998; Gunder and Krauth 1998; Wu and Huang 2010; Zhang et al. 2011a).

However, alongside with these advantages, MBR technology is affected by crucial issues that seriously hamper the performance and the widespread applications of MBRs (Bouhabila et al. 2001; Li et al. 2013; Mannina and Cosenza 2013; Zhang et al. 2011a). Membrane fouling, that is the undesirable deposition of retained particles, colloids, macromolecules and salts on the membrane surface or on the membrane pores, is the most critical disadvantage (Houari et al. 2010; Meng et al. 2009; Rana and Matsuura 2010; Kochkodan 2012). Specifically, membrane fouling results in a reduction of separation process output, diminishes process productivity, severe decline of the permeation flux or rapid transmembrane pressure increase (TMP), leading to high energy consumption, frequent membrane cleaning or replacement, which consequently leads to the increase in operating and maintenance cost (Asatekin et al. 2007; Bouhabila et al. 2001; Huang et al. 2010, 2012; Khan et al. 2009; Liao et al. 2004; Pan et al. 2009; Porcelli and Judd 2010; Rabie et al. 2001; Tansel et al. 2000; Wang et al. 2009).

In MBR, fouling mechanism is strongly related to physical and chemical interactions between the foulants and the membrane surface. Le-Clech et al. (2006) stated that the mechanism of fouling in MBRs includes a mixture of pore closure, cake layer and pore blocking, however, Zhang et al. (2006) argued that fouling mechanism involves three stages: adsorption, pore closure followed by a period of slow of TMP rise, and TMP jump. Chang et al. (2002) and Le-Clech et al. (2006) reviewed membrane fouling by focusing on factors that affect MBR fouling. They provided a very comprehensive study outlining numerous factors such as sludge characteristics, operational parameters, membrane materials and feed water characteristics which all interact in causing fouling in MBRs.

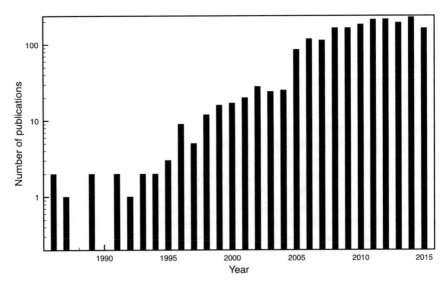

Fig. 1.4 Publications on MBR fouling (as reported in Scopus using the search term "membrane AND bioreactor AND fouling" for title-abstract-keyword on 25 August 2015)

Recently, numerous research studies have been conducted exploring fouling in MBRs and techniques used to overcome fouling. As can be seen in Fig. 1.4, a plethora of articles have been recently published on fouling and techniques to reduce its impact on MBR. Despite the extensive research activity related to MBR's, the concept of fouling and techniques to overcome fouling in MBR's is still an active topic, which is extensively discussed in the literature.

Apart from membrane fouling, cost is another major limitation of MBR technology. This is because of the high cost of membranes, which leads to the increases in both, operational and maintenance costs. Membrane cost covers replacing severely fouled or corrupted membranes and membrane cleaning processes during maintenance (Judd 2006; Le-Clech et al. 2006).

To sum up, various processes have been developed for water treatment using membranes. Among these processes, MBR stands out as one of the most efficient technologies. Its capability in treating wastewater cannot be disputed. However alongside their advantages, MBR's are limited by the concept of fouling. Membrane fouling is a major issue in the membrane bioreactors that significantly impair their practical applications. Therefore, this issue needs to be addressed to minimise the need for chemical cleaning and physical cleaning, reducing operational and maintenance costs that have a significant impact on MBRs. By minimising membrane fouling, membrane bioreactors will be more widely applicable for a range of wastewater treatment.

References

Asatekin A, Kang S, Elimelech M, Mayes AM (2007) Anti-fouling ultrafiltration membranes containing polyacrylonitrile-graft-poly(ethylene oxide) comb copolymer additives. J Membr Sci 298(1–2):136–146

Baker RW (2004) Membrane technology and applications, vol, 2nd edn. John Wiley Sons Ltd, England

Bella D, Durante F, Torregrossa M, Viviani G, Mercurio P, Cicala A (2007) The role of fouling mechanisms in a membrane bioreactor. Water Sci Technol 55(8–9):455–464

Bouhabila E, Ben R, Buisson H (2001) Fouling characterization in membrane bioreactors. Sep Purif Technol 22–23:123–132

Chang IS, Clech PL, Jefferson B, Judd S (2002) Membrane fouling in membrane bioreactors for wastewater treatment. J Environ Eng 128(11):1018–1029

Chang JS, Chang CY, Chen AC, Erdei L, Vigneswaran S (2006) Long-term operation of submerged membrane bioreactor for the treatment of high strength acrylonitrile-butadiene-styrene (ABS) wastewater: effect of hydraulic retention time. Desalination 191(1–3):45–51

Chu H, Li X (2005) Membrane fouling in a membrane bioreactor (MBR): sludge Cake formation and fouling characteristics. Biotechnol Bioeng 90(3):323–331

Côté P, Buisson H, Praderie M (1998) Immersed membranes activated sludge process applied to the treatment of municipal wastewater. Water Sci Technol 38(4-5-5 pt 4):437–442

Crawford G, Thompson D, Lozier J, Daigger G, Fleischer E (2000) Membrane bioreactors-ADesigner's perspective. In: Paper presented at the 73rd annual water environment federation technical exposition and conference Anaheim, Oct 14–18

Davis W, Le M, Heath C (1998) Intensified membranes activated sludge process with submerged membrane microfiltration. Water Sci 38(4–5):421–428

Drews A (2010) Membrane fouling in membrane bioreactors-characterisation, contradictions, causes and cures. J Membr Sci 368:1–28

Enegess D, Togna A, Sutton P (2003) Membrane separation applications to biosystems for waste water treatment. Filtr Sep 40:14–17

Gunder B, Krauth K (1998) Replacement of secondary clarification by membrane separation—Results with plate and hollow fiber modules. Water Sci 38(4–5):383–393

Hirani ZM, DeCarolis JF, Adham SS, Jacangelo JG (2010) Peak flux performance and microbial removal by selected membrane bioreactor systems. Water Res 44(8):2431–2440

Hong SP, Bae TH, Tak TM, Hong S, Randall A (2002) Fouling control in activated sludge submerged hollow fiber membrane bioreactors. Desalination 143(3):219–228

Houari A, Seyer D, Couquard F, Kecili K, Démocrate C, Heim V, Di Martino P (2010) Characterization of the biofouling and cleaning efficiency of nanofiltration membranes. Biofouling 26(1):15–21

Huang X, Xiao K, Shen Y (2010) Recent advances in membrane bioreactor technology for wastewater treatment in China. Front Environ Sci Eng China 4(3):245–271

Huang J, Zhang K, Wang K, Xie Z, Ladewig B, Wang H (2012) Fabrication of polyethersulfone-mesoporous silica nanocomposite ultrafiltration membranes with antifouling properties. J Membr Sci 423–424:362–370

Hussain A, Al-Rawajfeh AE, Alsaraierh H (2010) Membrane bio reactors (MBR) in waste water treatment: a review of the recent patents. Recent Pat Biotechnol 4(1):65–80. doi:10.2174/187220810790069505

Hwang BK, Kim JH, Ahn CH, Lee CH, Song JY, Ra YH (2010) Effect of disintegrated sludge recycling on membrane permeability in a membrane bioreactor combined with a turbulent jet flow ozone contactor. Water Res 44(6):1833–1840

Judd S (2006) Principles and applications of membrane bioreactors in water and waste water treatment. Oxford, London

Judd S (2008) The status of membrane bioreactor technology. Trends Technol 26(2):109–116

Judd S, Judd C (2010) Principles and applications of membrane bioreactors for water and waste water treatment, 2nd edn. Oxford, London

Khan SJ, Visvanathan C, Jegatheesan V (2009) Prediction of membrane fouling in MBR systems using emerically estimated specific cake resistance. Bioresour Technol 100(23):6133–6136

Kimura K, Yamato N, Yamamura H, Watanabe Y (2005) Membrane fouling in pilot-scale membrane bioreactors (MBRs) treating municipal wastewater. Environ Sci Technol 39(16):6293–6299

Kochkodan, V, DJ. Johnson, N. Hilal (2014) Polymeric membranes: Surface modification for minimizing (bio)colloidal fouling. Advances in Colloid and Interface Science 206:116-140

Koros WJ, Ma YH, Shimidzu T (1996) Terminology for membranes and membrane processes (IUPAC Recommendations 1996). J Membr Sci 120:149–159

Le-Clech P, Chen V, Fane TAG (2006) Fouling in membrane bioreactors used in wastewater treatment. J Membr Sci 284(1–2):17–53

Leiknes TO (2010) Membrane Bioreactors. In: Membrane technology. Wiley-VCH Verlag GmbH & Co. KGaA, pp 193–227. doi:10.1002/9783527631407.ch7

Li X, Chu H (2003) Membrane bioreactor for the drinking water treatment of polluted surface water supplies. Water Resour 37:4781–4791

Li Z, Tian Y, Ding Y, Chen L, Wang H (2013) Fouling potential evaluation of soluble microbial products (SMP) with different membrane surfaces in a hybrid membrane bioreactor using worm reactor for sludge reduction. Bioresour Technol 140:111–119

Liao BQ, Bagley DM, Kraemer HE, Leppard GG, Liss SN (2004) A review of biofouling and its control in membrane separation bioreactors. Water Environ Res 76(5):425–436

Liao B, Kraemer J, Bagley D (2006) Anaerobic membrane bioreactors: applications and research directions. Crit Rev Enivormen Sci Technol 36(6):489–530

Mannina G, Cosenza A (2013) The fouling phenomenon in membrane bioreactors: assessment of different strategies for energy saving. J Membr Sci 444:332–344

Meng F, Chae SR, Drews A, Kraume M, Shin HS, Yang F (2009) Recent advances in membrane bioreactors (MBRs): membrane fouling and membrane material. Water Res 43(6):1489–1512

Mulder M (1997) Basic principles of membrane technology, 2nd edn. Kluwer Academic publishers, Dordrecht

Nath K (2008) Membrane separation processes. Prentice Hall of India Private Limited, New Delhi

Pan J, Sun Y, Huang C (2009) Characteristics of soluble microbial products in membrane bioreactor and its effect on membrane fouling. Desalination 250:778–780

Perry RH, Green DW (1997) Perry's chemical engineers' handbook, 7th edn. McGraw-Hill, New York

Porcelli N, Judd S (2010) Chemical cleaning of potable water membranes: a review. Sep Purif Technol 71(2):137–143

Rabie HR, Côté P, Adams N (2001) A method for assessing membrane fouling in pilot- and full-scale systems. Desalination 141(3):237–243

Radjenović J, Matošić M, Mijatović I, Petrović M, Barceló D (2008) Membrane bioreactor (MBR) as an advanced wastewater treatment technology. Handbook of environmental chemistry. Water Pollut 5:S2. doi:10.1007/698_5_093

Rana D, Matsuura T (2010) Surface modifications for antifouling membranes. Chem Rev 110(4): 2448–2471

Rosenberger S, Kraume M (2002) Filterability of activated sludge in membrane bioreactors. Desalination 146(1–3):373–379

Rosenberger S, Kubin K, Kraume M (2002) Rheology of activated sludge in membrane activation reactors. Chemie-Ingenieur-Technik 74(4):487–494+373

Smith C, DiGregorio D, Talcott R (1969) The use of ultrafiltration membranes for activated sludge separation. Paper presented at the 24th annual Purdue industrial waste conference

Stephenson T, Judd S, Jefferson B, Brindle K (2000) Membrane Bioreactors for waste water treatment. IWA Publishing, London

Tansel B, Bao WY, Tansel IN (2000) Characterization of fouling kinetics in ultrafiltration systems by resistances in series model. Desalination 129(1):7–14

Tchobanoglous G, Burton FL, Stensel HD (2004) Wastewater engineering treatment and reuse, 4th edn. Mc-Graw Hill, New York

Thomas H, Judd S, Murrer J (2000) Fouling characteristics of membrane filtration in membrane bioreactors. Membr Technol 122:10–13

Visvanathan C, Aim RB, Parameshwaran K (2000) Membrane separation bioreactors for waste water treatment. Crit Rev Environ Sci Technol 30:1–48

Wang Z, Wu Z, Mai S, Yang C, Wang X, An Y, Zhou Z (2008) Research and applications of membrane bioreactors in China: progress and prospect. Sep Purif Technol 62(2):249–263

Wang Z, Wu Z, Tang S (2009) Extracellular polymeric substances (EPS) properties and their effects on membrane fouling in a submerged membrane bioreactor. Water Res 43(9): 2504–2512

Wu J, Huang X (2010) Use of azonation to mitigate fouling in a long-term membrane bioreactor. Bioresour Technol 101:6019–6027

Yamamoto K, Hiasa M, Mohmood T, Matsuo T (1989) Direct solid-liquid separation using hollow fiber membrane in an activated-sludge aeration tank. Water Sci Technol 21:43–54

Yamato N, Kimura K, Miyoshi T, Watanabe Y (2006) Difference in membrane fouling in membran e bioreactors (MBRs) caused by membrane polymer materials. Membr Sci 280 (1–2): 911–919

Yang W, Cicek N, Ilg J (2006) State-of-the-art of membrane bioreactors: worldwide research and commercial applications in North America. J Membr Sci 270(1–2):201–211

Zhang Q (2009) Performance evaluation and characterization of an innovative membrane bioreactor in the treatment of waste water and removal of pharmaceuticals and pesticides. Doctoral dissertation, University of Cincinnati

Zhang J, Chua HC, Zhou J, Fane AG (2006) Factors affecting the membrane performance in
    submerged membrane bioreactors. J Membr Sci 284(1–2):54–66. doi:10.1016/j.memsci.2006.
    06.022

Zhang H, Gao J, Jiang T, Gao D, Zhang S, Li H, Yang F (2011a) A novel approach to evaluate the
    permeability of cake layer during cross-flow filtration in the flocculants added membrane
    bioreactor. Bioresour Technol 102:11121–11131

Zhang K, Wei P, Yao M, Field R, Cui Z (2011b) Effect of the bubbling regimes on the
    performance and energy cost of flat sheet MBRs. Desalination 283:221–226

Chapter 2
Fundamentals of Membrane Processes

In this chapter the common fundamentals of different membrane processes are described. In the first part a general description of the different membrane structures, such as porous and dense homogeneous or symmetric and asymmetric membranes, and their function is discussed. In the second part membrane materials such as inorganic, organic, and composite materials and their function in membrane bioreactors are described. Third, the preparation of synthetic membranes via the phase inversion method is described in detail. Finally, alternative techniques of membrane fabrication are presented.

2.1 Membrane Classification by Membrane Structure

Synthetic membranes display a broad range in their physical structure and the material they are made from (Strathmann 2011). They can be classified according to their morphology, as shown in Fig. 2.1.

The first group is dense homogeneous polymer membranes. Usually they are prepared (i) from solution by solvent evaporation only or (ii) by extrusion of the melted polymer (Nunes and Peinemann 2006). However, dense homogeneous membranes only have a practical usefulness when they are made from a highly permeable polymer such as silicone. Commonly, the permeate flow across the membrane is quite low, since a minimal thickness is required to grant the membrane mechanical stability. The majority of membranes currently are porous or consist of a dense top layer on a porous structure (Mulder 1984, 1997; Nunes and Peinemann 2001; Strathmann et al. 2006).

© Springer Nature Singapore Pte Ltd. 2017
B. Ladewig and M.N.Z. Al-Shaeli, *Fundamentals of Membrane Bioreactors,*
Springer Transactions in Civil and Environmental Engineering,
DOI 10.1007/978-981-10-2014-8_2

Fig. 2.1 Membrane
classifications according to
morphology. Adapted from
Nunes and Peinemann (2006)

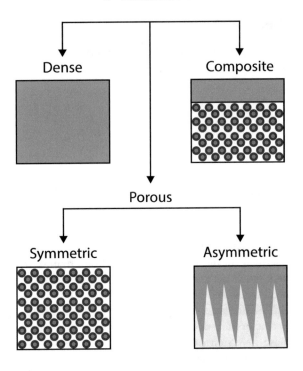

The second category is porous membranes, which can also be divided into two main groups. They are divided according to their pore diameter: microporous (d_p < 2 nm), mesoporous (2 nm < d_p < 50 nm) and macroporous (d_p > 50 nm) (Gallucci et al. 2011b).

The first groups of membranes are referred to as symmetric (isotropic) and the second type is referred to as asymmetric (anisotropic) membranes. Within the asymmetric membranes, there are several distinctly different structures including integrally skinned membranes (where the pore structure gradually changes from very large pores to very fine pores, essentially forming a "skin" on top of the membrane, giving rise to the name "integrally skinned"). Alternatively, the skin may be nonporous. A third, and industrially very important type of asymmetric membrane is the thin-film composite membrane, where a dense, selective, thin layer is deposited or polymerised at the surface/interface of a porous substrate.

Symmetric membranes refer to the membranes with uniform structure (uniform pore size or nonporous) throughout the entire membrane thickness (Buonomenna et al. 2011). Symmetric membranes are used today mainly in dialysis, electro-dialysis, and to some extent also in microfiltration (Strathmann 2000, 2011). The thickness of symmetric membranes is usually between 30 and 500 μm. The total resistance of the mass transfer relies on the total thickness of the membranes. Hence, a decrease in membrane thickness results in an increased permeation rate.

Asymmetric membranes have a gradient in structure. They consist of a 0.1–5 μm thick "skin" layer on a highly porous 100–300 μm thick structure (Strathmann 2011). The skin represents the actual selective barrier of the asymmetric sub-structure. Its separation properties are thoroughly determined by the nature of the material or the size of pores in the skin layer. The porous substrate layer serves as a support for the mostly very thin top layer, or "skin" (relatively dense) and has little effect on the separation properties of the membrane or the mass transfer rate of the membrane (Strathmann 2011). The dense surface layer is considered to be responsible for the membrane selectivity. Consequently, the controlled structure of the dense surface layer has become a serious concern in the membrane design (Zhenxin and Matsuura 1991). Also, the resistance to the mass transfer is mainly determined by the top layer (Buonomenna et al. 2011; Nunes and Peinemann 2001). Asymmetric membranes are primarily employed in pressure driven membrane processes such as reverse osmosis, ultrafiltration, gas separation and sometimes in microfiltration. High fluxes (high permeate flow per unit area), a reasonable mechanical stability providing very thin selective layer are the unique properties of asymmetric membranes (Strathmann 2011; Nunes and Peinemann 2006). Two procedures are used to prepare asymmetric membranes: the first method is based on phase inversion process which leads to integral structure (Kesting 1971), The second method resembles a composite structure in a two-step process in which a thin barrier layer is deposited on a microporous substructure (Cadotte and Petersen 1981; Strathmann 2011).

2.2 Membrane Materials

In general, there are three fundamentally different categories of membrane materials: Organic (polymeric), inorganic (ceramic) materials and biological materials. Organic materials are either cellulose—based or composed of modified organic polymers. By contrast, inorganic materials such as ceramics and metals are used in niche industrial applications but are usually cost-prohibitive in wastewater treatment. Biological membrane (bio membranes) is a selective barrier within or around a cell in a living organism. The biomembrane is capable of recognising what is necessary for the cell to receive or block for its survival. These membranes cannot meet the industrial requirements due to thermomechanical stability and productivity. It should be pointed out that, a large majority of membranes in research and commercial use are polymeric-based (organic membrane) as a result of their facile processing into viable membrane structures and the diverse polymers available, as well as the capability to synthesise novel polymer structures (Peyravi et al. 2012). Recently, composite membranes and inorganic membranes have gained tremendous attention owing to their potentially high performance, long lifetime and even their availability that outweigh the benefits/advantages of using polymeric membranes.

2.2.1 Inorganic Membranes

Inorganic membranes posses excellent thermal and chemical stability in comparison to polymeric membranes and hive higher antifouling property due to the hydrophilic nature of in organic material (Gallucci et al. 2011a; Mulder 1997). Nevertheless, there has been some limitation in their use despite their wide use and application. The main application of inorganic membranes in the past was enrichment of uranium hexafluoride U^{235} via Knudsen flow through porous ceramic membranes. Recently, many more applications are found in the field of ultrafiltration and microfiltration. Inorganic membranes are generally divided into four groups: glass membranes, ceramic membranes, metallic membranes, carbon membranes, and zeolitic membranes.

Metallic membranes can generally be obtained via the sintering of metal powders (e.g. stainless steel, molybdenum, or tungsten). According to (Gallucci et al. 2011a), the main materials for preparing metallic membranes are palladium and its alloys due to their high solubility and permeability of hydrogen. These membranes are employed for separation of hydrogen from gas mixtures and in the membrane reactor area for producing pure hydrogen (Lin 2001). These membranes have both advantages and disadvantages. The advantages are considerable mechanical strength and higher permeating flux (Gallucci et al. 2011a). The limitations of metallic membranes are (1) highly cost (very expensive) due to the low availability of palladium in nature and (2) surface poisoning, which is significantly more for thin metal membranes. There have been numerous studies reporting that the impact of poisons such as CO or H_2S on Pd-based membranes is a major problem. These gases (H_2S or CO) adsorb on the palladium surface that block the dissociation sites for hydrogen. Therefore, these membranes have received limited attention today because they do not relate to MBR technology (Judd 2006).

Ceramic membranes are of great importance in separation technology as they have a higher chemical, thermal and mechanical stability compared to organic membranes (Belfer et al. 2000). This stability makes these UF or MF membranes suitable in different fields of industry such as (food, biotechnology and pharmaceutical applications). They have been proposed for gas separation and the application of membrane reactor. Ceramic membranes are prepared through the combination of a metal (e.g. aluminium, titanium, silicium or zirconium, zinc, tin, and iron) with a non-metal in the form of oxide, nitride, or carbide to form a variety of inorganic nanoparticles (fillers) such as carbon nanotubes, alumina, or aluminium oxide (Al_2O_3), titanium oxide (TiO_2), zirconium dioxide or zirconia (ZrO_2), zinc oxide (ZnO), Silver, tin oxide (SnO_2), and Fe_3O_4. All these membranes have been used to fabricate inorganic—polymer composite membranes (Arthanareeswaran et al. 2008; Huang et al. 2006; Jian et al. 2006; Liang et al. 2012; Yang et al. 2007; Zoppi and Soares 2002; Leo et al. 2013; Celik and Choi 2011; Celik et al. 2011a, b; Bae and Tak 2005a, b; Maximous et al. 2010; Gallucci et al. 2011a; Leo et al. 2012; Lu et al. 2005; Mulder 1997; Rahimpour et al. 2008, 2009; Razmjou et al. 2011a, b; Sawada et al. 2012; Yu et al. 2009a, b; Livari et al. 2012; Moghimifar et al. 2014). Sintering or sol–gel techniques are usually used to prepare ceramic membranes.

Glass membranes can be regarded as ceramic membranes. Issues associated with ceramic membranes are the difficulties faced in proper sealing of the membranes in modules operating at high temperature, extremely high sensitivity of membranes to temperature gradient, leading to membrane cracking, and chemical instability of some perovskite-type materials (Gallucci et al. 2011a). Glass membranes (silica, SiO_2) are generally prepared by leaching techniques.

Carbon membranes (also called carbon molecular sieve membranes CMS) have been regarded as a promising candidate for applications of gas (Gallucci et al. 2011a). CMS are porous solids membranes, containing constricted holes that are responsible for approaching the molecular dimensions of diffusing gas molecules. Therefore, molecules with different size can be efficiently separated through molecular sieving (Gallucci et al. 2008). CMS membranes can be prepared by pyrolysis of thermosetting polymers such as poly acrylonitrile (PAN), cellulose triacetate, phenol formaldehyde, and poly (furfural) alcohol.

Recently, a new class of membranes have been developed and studied, such as the zeolitic membranes. These membranes have a very narrow pore size and can be employed in gas separation, pervaporation and separation of ions from aqueous solution by reverse osmosis. These membranes have some limitations; first the main limitation is relatively low gas fluxes in comparison to other inorganic membranes. Second, its thermal effect, as noted by Cejka et al. (2007), the zeolite layer exhibits negative thermal expansion, in which the zeolite layer shrinks when the region temperature is high, but the support layer expands continuously, causing thermal stress issues for the attachment of the zeolite layer to the support and for the connection of the individual microcrystals within the zeolite layer.

Inorganic membranes have both advantages and disadvantages as presented in Table 2.1. The major advantages of inorganic membranes when compared with polymeric membranes, is their high chemical, thermal, and mechanical stability and wide tolerance to pH (Belfer et al. 2000; Gallucci et al. 2011a). They can operate at high temperatures. As a fact of matter, inorganic membranes are stable at

Table 2.1 Advantages and disadvantages of inorganic membranes with respect to polymeric membranes (Gallucci et al. 2011a)

Advantages	Disadvantages
1. Long-term stability at high temperatures	1. High capital cost
2. Resistance to harsh environments (e.g. chemical degradation, pH, etc.)	2. Embrittlement phenomenon (in the case of dense Pd membranes)
3. Resistance to high pressure drops	3. Low membrane surface per module volume
4. Inertness to microbiological degradation	4. Difficulty of achieving high selectivities in large-scale microporous membranes
5. Easy cleanability after fouling	5. Low permeability of the highly hydrogen selective (dense) membranes at low temperatures
6. Easy catalytic activation	6. Difficult membrane to module sealing at high temperature

temperatures ranging from 300 to 800 °C and in some cases, ceramic membranes usable over 1000 °C (van Veen et al. 1996). They also have a high resistance to chemical degradation. Judd et al. (2004) stated that ceramic membranes did not foul substantially at fluxes up to 60 L m^{-2} h^{-1}, whilst polymeric membranes fouled at a lower flux of 36 L m^{-2} h^{-1}. Applicability of inorganic membranes is of great interest in non—aqueous filtration due to stability in organic solvents (Tsuru et al. 2000a, b). Despite their potential in waste water treatment, certain limitations deter membrane processes from large scale and continuous operation (Lee et al. 1999). One of the major limitations arises from membrane fouling caused by different inorganic salts (Bhattacharjee and Johnston 2002) which increases feed pressure, reduces permeate flux, decreases product quality and finally shortens membrane lifespan (Lee and Lee 2000; Seidel and Elimelech 2002).

Another limitation of inorganic membranes that probably hampers their application is the high capital costs of both the manufacturing process and material (Gallucci et al. 2011a). Therefore, the inorganic membranes might be used only in some special applications such as anaerobic biodegradation (Fan et al. 1996) and high temperature waste water treatment (e.g. high-strength industrial waste) (Luonsi et al. 2002; Scott et al. 1998). Despite, the high expense of inorganic membranes and their susceptibility to membrane fouling, they are still a competitive product in many applications. It is expected that inorganic membranes will have more applications in the future. Table 2.1 summarises the advantages and disadvantages of inorganic membranes over polymeric membranes.

2.2.2 Polymeric Membranes

Although polymer membranes are less resistance to high temperature and aggressive chemicals than inorganic or metallic membranes, they are still the most widely used materials in wastewater treatment applications. This is mainly owing to easy preparation, reasonable expense (low cost), high efficiency for removing dispersed oil, particles, and emulsified, small size, lower energy requirement, flexibility in membrane configuration, and relatively low operating temperature which is also associated with less stringent demands for the materials need in the construction of module (Buonomenna et al. 2011; Nunes and Peinemann 2010). Among many homopolymeric materials presented in Table 2.1, polyethersulfone (PES) is one of the most vital polymeric materials and is widely used in producing microfiltration (Li et al. 2008, 2009a, b; Ulbricht et al. 2007; Zhao et al. 2003a, b; Shin et al. 2005; Liu and Kim 2011), ultrafiltration (Chaturvedi et al. 2001; Marchese et al. 2003; Xu and Qusay 2004; Wang et al. 2006a, b, c) as well as nanofiltration membranes (Boussu et al. 2006; Ismail and Hassan 2007), either on the laboratory or industrial scale (Razali et al. 2013) Polyethersulfone (PES) has been recognised or acknowledged as a high performance polyaromatic polymer possessing toughness and thermal stability (Huang et al. 2012; Li et al. 2004; Shi et al. 2008; Shin et al. 2005; Zhao et al. 2013).

PES is a thermoplastic polymer and is typically amorphous in nature and shows one prominent XRD peak at $2\theta = 19.9°$ (Nair et al. 2001; Kumar et al. 2006; Guan et al. 2005). PES membranes show a high glass transition temperature ($T_g \approx 503$ K). PES structure has a harder benzene ring and a softer ether bond; so crystalline properties can be expected (Barth et al. 2000; Ismail and Hassan 2007). Additional properties include

- Good chemical resistance (inertness): PES exhibits excellent chemical resistance to aliphatic hydrocarbons, alcohols, and acids. It is also soluble in some aprotic polar solvents (Zhu et al. 2014).
- Blood compatibility.
- Outstanding oxidative stability.
- Outstandingly high mechanical strength.
- Easy processing and environmental endurance.
- Wide temperature and pH tolerance.
- Moderate good chlorine resistance.
- Easy to fabricate membranes in a wide variety of modules and configurations as well as wide range of pore size available for UF and MF from 10 A to 0.2 μm.
- PES also shows other good qualities such as good membrane forming properties.
- Commercially available and relatively inexpensive (Bolong et al. 2009).

Polyethersulfone has been affirmed as the membrane material in many processes such as in biomedical fields for blood purification (haemodialysis and plasma collection) (Barzin et al. 2004; Samtleben et al. 2003; Tullis et al. 2002; Werner et al. 1995; Zhao et al. 2001; Unger et al. 2005), hollow fibres (Khulbe et al. 2003) (Yang et al. 2007), stable substrate for the deposition and thermal processing of semiconductor thin films (Nair et al. 2001), sensors applications (Gerlach et al. 1998), sterilisation and pharmaceutical (Baker 2012; Song et al. 2000), water purification, beverage filtration, protein separation, and pre-treatment of reverse osmosis (Bolong et al. 2009; Yu et al. 2011).

However, its main limitations is related to the relatively high hydrophobic property, which can lead to severe membrane fouling (Kim et al. 1999; Van der Bruggen et al. 2002a, b) owing to the adsorption of nonpolar solutes and hydrophobic particles or bacteria onto its surface (Khulbe et al. 2000, 2010; Koh et al. 2005; Rahimpour and Madaeni 2010). This would lead to the gradual reduction of permeation flux, frequent membrane cleaning, impacting on the useable lifetime of the membrane and its applications (Daraei et al. 2013a, b, c; Luo et al. 2005; Yamamura et al. 2007a, b; Zhao et al. 2013; Rahimpour et al. 2011). Therefore, achieving the desired surface properties without modification of the advantageous properties of PES membrane is a paramount goal for membrane researchers and industry.

Regarding of the preparation of PES membranes, (Zhao et al. 2013) stated that the structure of PES membranes is always symmetric and is prepared by phase inversion methods. The structure of PES membrane is affected by the composition (e.g. concentration, solvent, and additives), temperature of PES solution, the non-solvent or the mixture of nonsolvents, and the coagulation bath or even the environment (Barth et al. 2000).

Many other polymeric materials can also be employed for fabricating membranes as explained schematically in Table 2.2. This table illustrates some examples of polymeric materials used for microfiltration, ultrafiltration, reverses osmosis, and some membrane processes.

Recently, copolymer is another important polymeric material in the manufacture of membranes. It is gaining more attention by a number of researchers. The copolymers are composed of at least two different types of structural units with different properties. The properties of copolymers rely on the properties of the units that are connected in the polymer and their relative proportions. Hence, polymers that are employed for the preparation of membrane and require different properties can be copolymerised carefully by selecting various polymeric units. To date, many membranes with high performance have been fabricated by different copolymers with different molecular structures (for example, block copolymers: poly (ethylene oxide)–polysulfone block copolymer (Hancock et al. 2000), graft copolymers: poly (ethylene glycol)-graft-polyacrylonitrile copolymer (Su et al. 2009b), and so on.

2.2.3 Composite Membranes

Composite membranes are often referred to as thin-film composite (TFC) membranes and they have received tremendous attention in recent years for desalination of brackish and sea water, waste water reclamation, and the separation and purification of chemical and biological products (Buch et al. 2008). TFC membranes are composed of at least two layers (with different (polymeric) materials), with a very selective membrane material being deposited as a dense ultrathin layer formed upon a more or less porous support layer (sublayer), which usually is an ultrafiltration membrane and serves as support as shown in Fig. 2.2 (Strathmann 1989; Wu 2012). The advantage of TFC membranes is each layer (i.e. top selective layer and bottom porous substrate) is thoroughly optimised and controlled independently to achieve the desired selectivity and permeability while presenting excellent compression resistance and mechanical strength (Jahanshahi et al. 2010; Lau et al. 2012; Rahimpour 2011; Kosaraju and Sirkar 2008). The porous support layer (bottom layer) is generally prepared through phase inversion method. On the other hand the top selective layer is prepared from elastomer, which is hard to prepare it through phase inversion method. The first generation TFC membranes were prepared by pouring a thin layer of polymer solution on a liquid of water or mercury (Mulder 1997). Numerous coating procedures have been used to prepare TFC membranes, including plasma polymerisation, dip coating, in situ polymerisation, and interfacial polymerisation. These techniques will be discussed in this chapter.

Table 2.2 Overview of polymeric materials used for membranes, the type of membranes formed and the membrane processes they are used in Ulbricht (2006)

Polymer	Barrier type	Morphology cross section	Barrier thickness (μm)	Membrane processes
Cellulose acetates	Nonporous	Anisotropic	~0.1	GS, RO
	Mesoporous	Anisotropic	~0.1	UF
	Macroporous	Isotropic	50–300	MF
Cellulose Nitrites	Macroporous	Isotropic	100–500	MF
Cellulose regenerated	Mesoporous	Anisotropic	~0.1	UF, D
Perfluorosulfonic acid polymer	Nonporous	Isotropic	50–500	ED, fuel cell
Polyacrylonitrile	Mesoporous	Anisotropic	~0.1	UF
Polyetherimides	Mesoporous	Anisotropic	~0.1	UF
Polyethersulfone	Mesoporous	Anisotropic	~0.1	UF
	Macroporous	Isotropic	50–300	MF
Polyethylene terephthalate	Macroporous	Isotropic track-etched	6–35	MF
Polyphenylene oxide	Nonporous	Anisotropic	~0.1	GS
Polytetrafluoroethylene	Macroporous	Isotropic	50–500	MF
	Nonporous		~0.1	GS
Polyamide, aliphatic	Macroporous	Isotropic	100–500	MF
Polyamide, aromatic	Mesoporous	Anisotropic	~0.1	UF
Polyamide, aromatic. In situ synthesised	Nonporous	Anisotropic/composite	~0.05	RO, NF
Polycarbonates, aromatic	Nonporous	Anisotropic	~0.1	MF
	Macroporous	Isotropic track-etched	6–35	RO, NF
Polyether, aliphatic crosslinked, In situ synthesised	Nonporous	Anisotropic/composite	~0.05	RO, NF
Polyethylene	Macroporous	Isotropic	50–500	MF
Polyimides	Nonporous	Anisotropic	~0.1	GS, NF
Polypropylene	Macroporous	Isotropic	50–500	MF

(continued)

Table 2.2 (continued)

Polymer	Barrier type	Morphology cross section	Barrier thickness (μm)	Membrane processes
Polysiloxanes	Nonporous	Anisotropic/composite	~0.1 < 1–10	GS, PV, NF (organophilic)
Polysulfone	Nonporous Mesoporous	Anisotropic Anisotropic	~0.1 ~0.1	GS UF
Polyvinyl alcohol, crosslinked	Nonporous	Anisotropic/composite	<1–10	PV (hydrophilic)
Polyvinylidene fluoride	Mesoporous Macroporous	Anisotropic Isotropic	~0.1 50–300	UF MF
Poly(Styrene-co-divinylbenzene), sulfonated or terminated	Nonporous	Isotropic	100–500	ED

RO reverse osmosis, *NF* nanofiltration, *UF* ultrafiltration, *MF* microfiltration, *GS* gas separation, *ED* electrodialysis, *PV* pervaporation

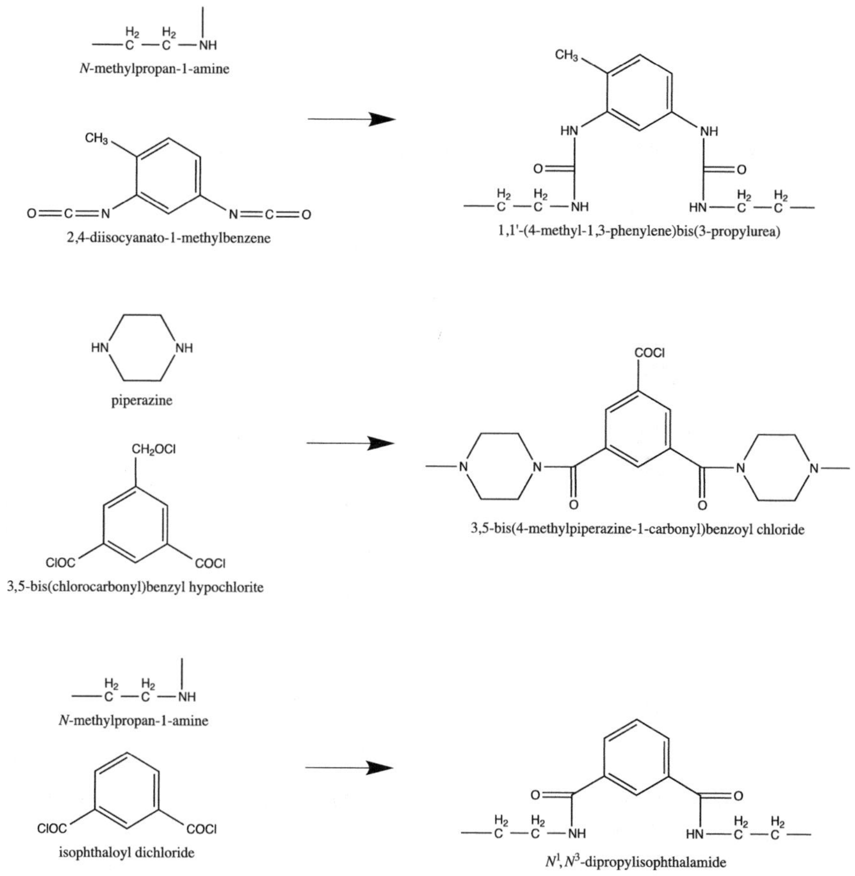

Fig. 2.2 Examples of the preparation of composite membranes by interfacial polymerization (Mulder 1997)

Another type of procedure used to prepare composite membranes, is the coating layer, which plugs the pores in the sublayer. In this procedure, the unique characteristics of the sublayer compared to those of the coating layer highlight the overall properties. With other techniques like stretching, sintering, leaching out and track-etching, porous membranes can be obtained. Porous membranes are mainly used as sublayer for composite membranes. Using phase inversion method is frequently possible to obtain open or dense structures. Various techniques have been implemented to prepare the ultrathin barrier layer upon the supporting layer. These techniques include (Lau et al. 2012; Mulder 1997; Seman et al. 2012).

(a) Interfacial polymerization (IP): It is a technique that is used for depositing the thin selective layer onto the porous layer. Polymerisation reactions emerge between two reactive monomers that react on the interface of two immiscible phase (an aqueous phase and an organic solvent such as hexane) through

interfacial polymerization, and form a denser polymeric top layer on the
supporting layer surface. The benefits of this technique are self-inhibiting
through passage of a limited supply of reactants through the already formed
layer, causing an extremely thin film of thickness in the range of 50 nm.
Figure 2.2 illustrates some examples of monomers and pre-polymers that can
be used to prepare composite membranes by interfacial polymerization.

(b) Dip coating: It is a straightforward and effective technique to prepare composite membranes with a very thin structure but dense toplayer. Membranes
obtained by this technique can be used in gas separation, pervaporation and
reverse osmosis applications. The principle of this method is dip coating a
polymer solution onto the supporting layer's surface and then drying the
coated layer (using oven).

(c) Lamination: In this technique, casting an ultrathin layer and then the microporous layer is covered with the casted ultrathin layer.

(d) Plasma-initiated polymerization: This technique involves depositing the barrier
film directly on the microporous support layer by gaseous phase monomer
plasma using an electrical discharge at high frequencies up to 10 MHz (Mulder
1997).

2.2.4 Formation of Membranes by Phase Inversion Method

Generally, various techniques have been extensively used to prepare synthetic
membranes either inorganic membranes (e.g. ceramics, metals, glass, Zeolites) or
organic membranes that includes all sorts of polymers (Mulder 1997; Nunes and
Peinemann 2006; Strathmann et al. 2006). The purpose of preparation is to modify
the material using an appropriate technique to obtain a membrane structure with an
appropriate morphology for a specific separation. The preparation method is limited
by the materials used, the membrane morphology obtained and the separation
principle applied.

The techniques that are being employed for the preparation of synthetic membrane are phase inversion, stretching of films, irradiation and etching of films,
sintering of powders, track-etching, sol–gel process, microfabrication vapour
deposition and coating (Hoek and Elimelech 2003; Hoek et al. 2002; Jeong et al.
2007; Mulder 1997; Tang et al. 2008; Ulbricht 2006). One of the most important
methods is phase inversion.

Phase inversion method is arguably one of the most common and versatile
technique used to prepare all sorts of morphologies (both symmetric and asymmetric types) due to the significance of immersion precipitation (Boom et al. 1992;
Buonomenna et al. 2011; Madaeni and Rahimpour 2005a, b; Mulder 1997;
Rahimpour and Madaeni 2010; Rahimpour et al. 2007a, b, 2009, 2010a, b). It is a
process whereby a polymer is transformed in a controlled manner from a liquid
state to a solid state. Throughout this technique, a thermodynamically stable

polymer solution with a multiple components is subjected to a liquid–liquid demixing whereby the cast polymer film separates into a polymer—rich phase (membrane matrix) and a polymer—lean phase (membrane pores) (Buonomenna et al. 2011). The mechanism of phase inversion during membrane formation can be concisely described by a polymer/solvent/nonsolvent system, as explained typically by a ternary phase diagram shown in Fig. 2.3 (Buonomenna et al. 2011).

The corners of the triangle act the three components (polymer, solvent, non-solvent), whilst at any point within the triangle represents a mixture of three components (starting compositions of the casting solution); the binodal curve divides the triangle into two phase regions; a one-phase region where all components are miscible to and a two-phase region where the system is separated into a polymer-rich, generally a solid phase and a polymer-poor which is generally liquid phase (although the one-phase region in the phase diagram is continuous thermodynamically).

For practical purposes it can conveniently be divided into a liquid and a solid region. The tie lines within the two-phase region connect two equilibrium states on the binodal curve, which also represent the compositions of two coexisting phases generated during the phase separation. The region between the spinodal and the binodal curves is called metastable region, where phase separation appears under certain initiation (nucleation). The region within the spinodal curve corresponds to unstable composition where immediate demixing occurs after entering this region.

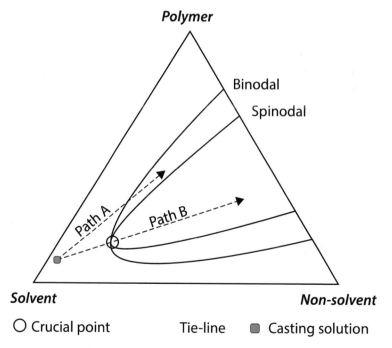

Fig. 2.3 Ternary phase diagram (solvent/polymer/nonsolvent) for membrane formation via phase inversion process (Machado et al. 1999)

By immersing a thin layer casting solution into a coagulation bath, the solvent of the casting solution is exchanged with a nonsolvent (Rahimpour and Madaeni 2007). The composition of the coagulation and casting solution are the most considerable factor, as it determines the phase inversion path of a membrane forming system (Albrecht et al. 2001).

The filled square represents the initial state of casting solution. The exchange between solvent and nonsolvent changes the composition in the casting film. Once demixing polymer solution arrives in the metastable region between the binodal and the spinodal, the region is referred to "binodal demixing" and therefore represents path A (Fig. 2.3). In this region, the polymer solution is separate into a polymer-lean phase and a polymer-rich phase (Buonomenna et al. 2011). Another pathway towards miscibility gap (path B) is called "spinodal decomposition". In this pathway, the composition path passed through the thermodynamically unstable zone (critical point), in which the binodal and spinodal curve converge, and two co-continuous phases formed. This process yields asymmetric membranes with a dense top layer and porous sublayer containing macrovoids, pores, and micropores (Rahimpour and Madaeni 2007).

The general concept of phase inversion method covers a variety of different techniques, including:

(a) Precipitation by solvent evaporation: Precipitation by solvent evaporation is a straightforward technique used to prepare phase inversion membranes. The polymer is dissolved in a solvent and form homogenous solution (casting or dope solution). Then, the dope solution is casted on a suitable support (e.g. glass plate) or another type of support that may be nonporous (polymer) or porous (non-woven fabric). Later, the solvent is allowed to evaporate in an inert gas (e.g. nitrogen) to exclude water vapour, allowing a dense homogenous membrane to be obtained (Mulder 1997).

(b) Precipitation by controlled evaporation: In the early years this technique has been used. In this method, the polymer is dissolved in a mixture of solvent and nonsolvent. Even though, the solvent is more volatile than the nonsolvent, the evaporation step leads to polymer precipitation own to higher nonsolvent content. The structure of membrane prepared by this technique is skinned membranes.

(c) Thermal precipitation: Polymeric solution in a mixed or single solvent is cooled to allow phase separation to occur. Evaporation of the solvent usually allows the formation of a skinned membrane (Cheryan 1998; Oh et al. 2009; Su et al. 2009a, b). This method is thoroughly used to prepare microfiltration membranes.

(d) Precipitation from the vapour phase: In this method, a dope solution, which consists of a polymer and a solvent, is placed in a vapour atmosphere whereas vapour phase consists of a nonsolvent saturated with the same solvent. The high solvent concentration in the vapour phase prohibits the evaporation of solvent from the cast film. As a result of diffusion of nonsolvent into the cast film, the membrane formation appears. This leads to formation of porous membrane without a toplayer.

(e) Immersion precipitation or nonsolvent induced phase inversion: This is one of the most common methods in the preparation of the most polymeric microfiltration and ultrafiltration and some of nanofiltration which is used for separation processes (Ismail and Hassan 2007). In this procedure, a film of homogenous polymeric solution is cast on a suitable substrate after preparing it by dissolving polymer into solvent. Then the cast film is immersed in a coagulation bath containing deionised water or methanol. Precipitation occurs as a consequence of the exchange between the solvent and nonsolvent. The membrane structure essentially is obtained from a combination of mass transfer and phase separation method (Rahimpour et al. 2010a). Keeping in mind, that thermodynamic behaviour of a polymer solution is attributed to immersion—precipitation and is represented by polymer/solvent/nonsolvent systems (Buonomenna et al. 2011). This precipitation is also called dry/wet method (Zhao et al. 2008; Oh et al. 2009).

2.3 Membrane Fabrication Techniques

Membranes of different arrangements are formed in order to meet different industrial and domestic demands; thus, various casting techniques have been implemented to obtain different types of the membrane such as flat sheet membrane (casting), hollow fibre membrane (spinning) and composite membrane (dip coating). The first two, will be discussed in this chapter. The third one is discussed previously.

2.3.1 Flat Sheet Membranes

For research purposes, flat sheet membrane is a relatively simple method used to fabricate/prepare membranes. In industrial scale, the casting method employed is usually a continuous mode as shown below in Fig. 2.4 (Mulder 1997).

The principle of this method is the polymer is initially dissolved in an appropriate solvent mixture (which may include additives) and forms a homogeneous (dope) solution. Molecular weight of a polymer, concentration of the polymer and the kind of solvent used (mixture) are three factors (parameters) affect the viscosity of the dope solution. Afterward, the polymer solution is spread (poured) and cast directly to a thin film of a homogenous polymer solution using one of the supporting layer (for example, clean glass plate, or polyethylene non-woven fabric, polyester, metal, and Teflon) by means of steel casting knife and adjusting the thickness of the membranes. The casting thickness can roughly vary from 50 to 500 μm. The thin film of a homogenous polymer solution (protomembrane) is immediately immersed in a second liquid, which is a nonsolvent for the polymer;

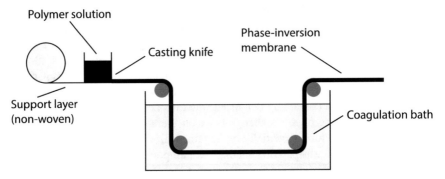

Fig. 2.4 Schematic of continuous flat sheet membrane preparation. Adapted from Mulder (1984, 1997)

however, it is miscible with the polymer solvent. Exchange between the solvent and nonsolvent occurs and ultimately introduces phase separation in polymer film and would lead to the formation of membranes (Rahimpour and Madaeni 2010; Rahimpour et al. 2010a, b; Mulder 1997). Water is used as a nonsolvent (second liquid), as it is a powerful nonsolvent. Organic solvents (e.g. methanol) can be used for the same purpose as well. Since the solvent/nonsolvent pair is a very important characteristic in obtaining the desired structure, the nonsolvent cannot be chosen at will.

A non-continuous mode is usually employed in a laboratory scale. The three-component polymer solution (polymer/solvent/additive) is prepared and stirred under particular temperatures in order to ensure complete dissolution of the polymer. After the polymer solution is placed for some time and the complete release of bubbles is confirmed, the homogenous solution is cast on glass plates using a casting knife with a specific thickness (for example, filmographe Dr. Blade 150 μm; Erichsen blade 150 μm). This is immediately moved to the coagulation bath (which is usually deionised water) for immersion at room temperature without any evaporation. Then, the membranes are peeled off the glass and subsequently rinsed with deionised water and stored in fresh deionised water for at least one day to leach out all residual solvents (Rahimpour and Madaeni 2010; Rahimpour et al. 2009, 2010a, b). At the final stage, membrane is sandwiched by placing between two sheets of filter paper or placing in air for 24 h at room temperature.

In Summary, flat sheet membranes are relatively straightforward to prepare, as they are very effective for characterising on laboratory scale. A dead end cell station is usually used for measuring water flux of membranes.

For very small membrane surface area (less than 1000 cm^2), the membranes are mostly cast by hand or semi-automatically using glass plate, not on non-woven polyester.

2.3.2 Hollow Fibre Membranes

Spinning is another technique used to prepare hollow fibre membranes. For industrial applications, hollow fibre membranes are more applicable, more effective and cheaper than flat sheet membranes because hollow fibre membranes have relatively higher (surface/volume) ratio compared to flat sheet membranes, which can have a greater resistance to pressure. Additionally very few manufactures supply flat sheet membranes (Zhao et al. 2013). Hollow fibre membranes have a minimum dead space; therefore it can be physically cleaned by frequent backwash, yielding longer life to the membranes. It has been designed with specific dimensions that are suitable to minimise membrane fouling in a given application. Many scholars believe that hollow fibre membranes and flat membranes can exhibit similar performance. But, the procedures for their preparation are thoroughly different, considering hollow fibre membranes are self-supporting. The fibre dimensions are a paramount aspect and should be taken into account when preparing hollow fibre membranes.

In general, hollow fibre membranes can be prepared by three methods:

(a) Wet spinning method: possible but rarely used to prepare hollow fibre membranes,
(b) Melt spinning, and
(c) Dry spinning (dry-wet spinning): this method is more applicable to prepare hollow fibre membranes and is based on the phase inversion method (Khayet and García-Payo 2009).

A schematic drawing illustrating the preparation of hollow fibre membranes is displayed in Fig. 2.5. The polymer solution is prepared and stored in a thermostated tank. Then, the solution is pumped and extruded through a tube–in-orifice spinneret; the polymer solution (being filtered before it) enters the spinneret. The viscosity of the polymer solution should be high (more than 100 poise). The bore injection liquid (nonsolvent liquid or gas) is also delivered or passed through the inner tube of the spinneret. The primary function (goal) of the bore liquid is to keep the fibre open and to assist in controlling the interior surface morphology of the hollow fibre through phase inversion.

After a short residence time in air or atmosphere, the fibre is immersed in a coagulation bath where precipitation appears outside the liquid filament due to solvent evaporation. After immediate coagulation, asymmetric hollow fibre are formed with density gradient along the radial direction (Machado et al. 1999). The fibres are then rinsed and cut with the desired length and collected upon a godet. Cutting the fibre is favourable in the hollow fibre process to assist in promoting the flow of bore liquid inside the hollow. The main parameters in the spinning technique are (Mulder 1997):

(a) The extrusion rate of the polymer solution,
(b) The bore fluid rate,
(c) The tearing rate,

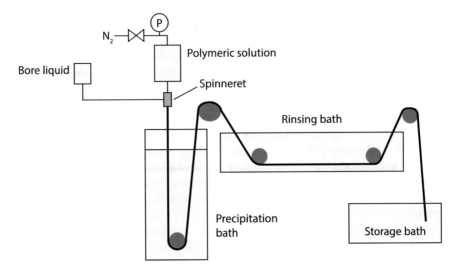

Fig. 2.5 Hollow fibre spinning apparatus. Adapted from Machado et al. (1999), Mulder (1997)

(d) The residence time in the air gap (between spinneret and coagulation bath), and
(e) The dimensions of the spinneret.

These parameters strongly interact with the membrane forming parameters such as the composition of the polymer solution, the composition of the coagulation bath and its temperature.

In summary, the configurations of flat sheet and hollow fibre membrane can be used as membrane bioreactors (MBRs) for applications of wastewater treatment, and both have their advantages and disadvantages.

References

Albrecht W, Weigel T, Schossig-Tiedemann M, Kneifel K, Peinemann KV, Paul D (2001) Formation of hollow fiber membranes from poly(ether imide) at wet phase inversion using binary mixtures of solvents for the preparation of the dope. J Membr Sci 192(1–2):217–230. doi:10.1016/S0376-7388(01)00504-X

Arthanareeswaran G, Sriyamuna Devi TK, Raajenthiren M (2008) Effect of silica particles on cellulose acetate blend ultrafiltration membranes: Part I. Sep Purif Technol 64(1):38–47

Bae TH, Tak TM (2005a) Effect of TiO_2 nanoparticles on fouling mitigation of ultrafiltration membranes for activated sludge filtration. J Membr Sci 249(1–2):1–8

Bae TH, Tak TM (2005b) Interpretation of fouling characteristics of ultrafiltration membranes during the filtration of membrane bioreactor mixed liquor. J Membr Sci 264(1–2):151–160

Baker R (2012) Membrane technology and application, 3rd edn. Wiley, United Kingdom

Barth C, Gonçalves MC, Pires ATN, Roeder J, Wolf BA (2000) Asymmetric polysulfone and polyethersulfone membranes: effects of thermodynamic conditions during formation on their performance. J Membr Sci 169(2):287–299

Barzin J, Feng C, Khulbe KC, Matsuura T, Madaeni SS, Mirzadeh H (2004) Characterization of polyethersulfone hemodialysis membrane by ultrafiltration and atomic force microscopy. J Membr Sci 237(1–2):77–85

Belfer S, Fainchtain R, Purinson Y, Kedem O (2000) Surface characterization by FTIR-ATR spectroscopy of polyethersulfone membranes-unmodified, modified and protein fouled. J Membr Sci 172(1–2):113–124

Bhattacharjee S, Johnston GM (2002) A model of membrane fouling by salt precipitation from multicomponent ionic mixtures in crossflow nanofiltration. Environ Eng Sci 19(6):399–412

Bolong N, Ismail AF, Salim MR, Rana D, Matsuura T (2009) Development and characterization of novel charged surface modification macromolecule to polyethersulfone hollow fiber membrane with polyvinylpyrrolidone and water. J Membr Sci 331(1–2):40–49

Boom RM, Wienk IM, Den Van, Boomgaard T, Smolders CA (1992) Microstructures in phase inversion membranes. Part 2. The role of a polymeric additive. J Membr Sci 73(2–3):277–292

Boussu K, Van der Bruggen B, Volodin A, Van Haesendonck C, Delcour JA, Van der Meeren P, Vandecasteele C (2006) Characterization of commercial nanofiltration membranes and comparison with self-made polyethersulfone membranes. Desalination 191(1–3):245–253

Buch PR, Jagan Mohan D, Reddy AVR (2008) Preparation, characterization and chlorine stability of aromatic-cycloaliphatic polyamide thin film composite membranes. J Membr Sci 309(1–2):36–44

Buonomenna M, Choi S, Galiano F, Drioli E (2011) Membranes prepared via phase inversion. In: Basile A, Gallucci F (eds) Membranes for membrane reactors, preparation, optimization and selection, 1st edn. Wiley, UK, pp 475–490

Cadotte J, Petersen R (1981) Thin film reverse osmosis membranes: origin, development, and recent advances. In: Synthetic membranes, ACS symposium series 153, vol 1, pp 305–325

Cejka J, Corna HV, Corma A, Schuth F (2007) Introduction to zeolite science and practise, studies in surface science and catalysis. Elsevier

Celik E, Choi H (2011) Carbon nanotube/polyethersulfone composite membranes for water filtration. In: ACS Symposium Series, vol 1078

Celik E, Liu L, Choi H (2011a) Protein fouling behavior of carbon nanotube/polyethersulfone composite membranes during water filtration. Water Res 45(16):5287–5294

Celik E, Park H, Choi H, Choi H (2011b) Carbon nanotube blended polyethersulfone membranes for fouling control in water treatment. Water Res 45(1):274–282

Chaturvedi BK, Ghosh AK, Ramachandhran V, Trivedi MK, Hanra MS, Misra BM (2001) Preparation, characterization and performance of polyethersulfone ultrafiltration membranes. Desalination 133(1):31–40

Cheryan M (1998) Ultrafiltration and microfiltration handbook. Technomic Publishing Company Inc, USA

Daraei P, Madaeni SS, Ghaemi N, Khadivi MA, Astinchap B, Moradian R (2013a) Enhancing antifouling capability of PES membrane via mixing with various types of polymer modified multi-walled carbon nanotube. J Membr Sci 444:184–191

Daraei P, Madaeni SS, Ghaemi N, Khadivi MA, Astinchap B, Moradian R (2013b) Fouling resistant mixed matrix polyethersulfone membranes blended with magnetic nanoparticles: study of magnetic field induced casting. Sep Purif Technol 109:111–121

Daraei P, Madaeni SS, Ghaemi N, Khadivi MA, Rajabi L, Derakhshan AA, Seyedpour F (2013c) PAA grafting onto new acrylate-alumoxane/PES mixed matrix nano-enhanced membrane: preparation, characterization and performance in dye removal. Chem Eng J 221:111–123

Fan XJ, Urbain V, Qian Y, Manem J (1996) Nitrification and mass balance with a membrane bioreactor for municipal wastewater treatment. Water Sci Technol 34. doi:10.1016/0273-1223 (96)00502-1

Gallucci F, Basile A, Hai F (2011a) Membranes for membrane reactors: preparation, optimization and selection, 1st edn. Wiley, UK

Gallucci F, Basile A, Hai FI (2011b) Introduction—a review of membrane reactors. In: Membranes for membrane reactors: preparation, optimization and selection. Wiley, Chichester, pp 1–61. doi:10.1002/9780470977569.ch

Gallucci F, Tosti S, Basile A (2008) Pd-Ag tubular membrane reactors for methane dry reforming: a reactive method for CO_2 consumption and H_2 production. J Membr Sci 317(1–2):96–105. doi:10.1016/j.memsci.2007.03.058

Gerlach G, Baumann K, Buchhold R, Naklal A (1998) German Patent D E 19853732

Guan R, Zou H, Lu D, Gong C, Liu Y (2005) Polyethersulfone sulfonated by chlorosulfonic acid and its membrane characteristics. Eur Polym J 41(7):1554–1560

Hancock LF, Fagan SM, Ziolo MS (2000) Hydrophilic, semipermeable membranes fabricated with poly(ethylene oxide)-polysulfone block copolymer. Biomaterials 21(7):725–733

Hoek EMV, Elimelech M (2003) Cake-enhanced concentration polarization: a new fouling mechanism for salt-rejecting membranes. Environ Sci Technol 37(24):5581–5588

Hoek EMV, Kim AS, Elimelech M (2002) Influence of crossflow membrane filter geometry and shear rate on colloidal fouling in reverse osmosis and nanofiltration separations. Environ Eng Sci 19(6):357–372

Huang J, Zhang K, Wang K, Xie Z, Ladewig B, Wang H (2012) Fabrication of polyethersulfone-mesoporous silica nanocomposite ultrafiltration membranes with antifouling properties. J Membr Sci 423–424:362–370

Huang ZQ, Chen ZY, Guo XP, Zhang Z, Guo CL (2006) Structures and separation properties of PAN-Fe3O4 ultrafiltration membranes prepared under an orthogonal magnetic field. Ind Eng Chem Res 45(23):7905–7912

Ismail AF, Hassan AR (2007) Effect of additive contents on the performances and structural properties of asymmetric polyethersulfone (PES) nanofiltration membranes. Sep Purif Technol 55(1):98–109

Jahanshahi M, Rahimpour A, Peyravi M (2010) Developing thin film composite poly (piperazine-amide) and poly(vinyl-alcohol) nanofiltration membranes. Desalination 257(1–3):129–136

Jeong BH, Hoek EMV, Yan Y, Subramani A, Huang X, Hurwitz G, Ghosh AK, Jawor A (2007) Interfacial polymerization of thin film nanocomposites: a new concept for reverse osmosis membranes. J Membr Sci 294(1–2):1–7

Jian P, Yahui H, Yang W, Linlin L (2006) Preparation of polysulfone-Fe_3O_4 composite ultrafiltration membrane and its behavior in magnetic field. J Membr Sci 284(1–2):9–16

Judd S (2006) Principles and applications of membrane bioreactors in water and waste water treatment, 1st edn. Elsevier, UK

Judd S, Robinson R, Holdner J, Vazquez HA, Jefferson B (2004) Impact of membrane material on membrane bioreactor permeability. In: Proceedings of the water environment-membrane technology conference, Seoul, Korea

Kesting R (1971) Synthetic polymeric membranes. McGraw-Hill, New York

Khayet M, García-Payo MC (2009) X-Ray diffraction study of polyethersulfone polymer, flat-sheet and hollow fibers prepared from the same under different gas-gaps. Desalination 245 (1–3):494–500

Khulbe KC, Feng C, Matsuura T, Kapantaidakis GC, Wessling M, Koops GH (2003) Characterization of polyethersulfone-polyimide hollow fiber membranes by atomic force microscopy and contact angle goniometery. J Membr Sci 226(1–2):63–73. doi:10.1016/j. memsci.2003.08.011

Khulbe KC, Feng CY, Matsuura T (2010) Surface modification of synthetic polymeric membranes for filtration and gas separation. Recent Pat Chem Eng 3(1):1–16

Khulbe KC, Matsuura T, Singh S, Lamarche G, Noh SH (2000) Study on fouling of ultrafiltration membrane by electron spin resonance. J Membr Sci 167(2):263–273

Kim IC, Choi JG, Tak TM (1999) Sulfonated polyethersulfone by heterogeneous method and its membrane performances. J Appl Polym Sci 74(8):2046–2055

Koh M, Clark MM, Howe KJ (2005) Filtration of lake natural organic matter: adsorption capacity of a polypropylene microfilter. J Membr Sci 256(1–2):169–175

Kosaraju P, Sirkar K (2008) Interfacially polymerized thin film comopsite membranes on microporous polypropylene supports for solvent-resistant nanofiltration. Membr Sci 321:155–161

Kumar R et al (2006) Nucl Instr Meth Phys Res 248B: 279–283

Lau WJ, Ismail AF, Misdan N, Kassim MA (2012) A recent progress in thin film composite membrane: a review. Desalination 287:190–199

Lee S, Kim J, Lee CH (1999) Analysis of $CaSO_4$ scale formation mechanism in various nanofiltration modules. J Membr Sci 163(1):63–74

Lee S, Lee CH (2000) Effect of operating conditions on $CaSO_4$ scale formation mechanism in nanofiltration for water softening. Water Res 34(15):3854–3866

Leo CP, Ahmad Kamil NH, Junaidi MUM, Kamal SNM, Ahmad AL (2013) The potential of SAPO-44 zeolite filler in fouling mitigation of polysulfone ultrafiltration membrane. Sep Purif Technol 103:84–91

Leo CP, Cathie Lee WP, Ahmad AL, Mohammad AW (2012) Polysulfone membranes blended with ZnO nanoparticles for reducing fouling by oleic acid. Sep Purif Technol 89:51–56

Li B, Zhao W, Su Y, Jiang Z, Dong X, Liu W (2009a) Enhanced desulfurization performance and swelling resistance of asymmetric hydrophilic pervaporation membrane prepared through surface segregation technique. J Membr Sci 326(2):556–563

Li JF, Xu ZL, Yang H, Yu LY, Liu M (2009b) Effect of TiO_2 nanoparticles on the surface morphology and performance of microporous PES membrane. Appl Surf Sci 255(9): 4725–4732

Li Y, Cao C, Chung TS, Pramoda KP (2004) Fabrication of dual-layer polyethersulfone (PES) hollow fiber membranes with an ultrathin dense-selective layer for gas separation. J Membr Sci 245(1–2):53–60

Li Y, Chung TS, Xiao Y (2008) Superior gas separation performance of dual-layer hollow fiber membranes with an ultrathin dense-selective layer. J Membr Sci 325(1):23–27

Liang S, Xiao K, Mo Y, Huang X (2012) A novel ZnO nanoparticle blended polyvinylidene fluoride membrane for anti-irreversible fouling. J Membr Sci 394–395:184–192

Lin YS (2001) Microporous and dense inorganic membranes: current status and prospective. Sep Purif Technol 25(1–3):39–55. doi:10.1016/S1383-5866(01)00089-2

Liu SX, Kim JT (2011) Characterization of surface modification of polyethersulfone membrane. J Adhes Sci Technol 25(1–3):193–212

Livari AE, Aroujalian A, Raisi A, Fathizadeh M (2012) The effect of TiO_2 nanoparticles on PES UF membrane fouling in water oil separation. Process Eng 44:1783–1785

Lu Y, Yu S, Chai B (2005) Preparation of PVDF ultrafiltration membrane modified by nano-sized alumina(Al2O$_3$) and its antifouling research. Polymer 46:7701–7706

Luo ML, Zhao JQ, Tang W, Pu CS (2005) Hydrophilic modification of poly(ether sulfone) ultrafiltration membrane surface by self-assembly of TiO_2 nanoparticles. Appl Surf Sci 249(1–4):76–84

Luonsi A, Laitinen N, Beyer K, Levänen E, Poussade Y, Nyström M (2002) Separation of CTMP mill-activated sludge with ceramic membranes. Desalination 146(1–3):399–404

Machado PST, Habert AC, Borges CP (1999) Membrane formation mechanism based on precipitation kinetics and membrane morphology: Flat and hollow fiber polysulfone membranes. J Membr Sci 155(2):171–183

Madaeni SS, Rahimpour A (2005a) Effect of type of solvent and non-solvents on morphology and performance of polysulfone and polyethersulfone ultrafiltration membranes for milk concentration. Polym Adv Technol 16(10):717–724

Madaeni SS, Rahimpour A (2005b) Preparation of polyethersulfone ultrafiltration membranes for milk concentration and effects of additives on their morphology and performance. Chin J Polym Sci (English Edition) 23(5):539–548

Marchese J, Ponce M, Ochoa NA, Prádanos P, Palacio L, Hernández A (2003) Fouling behaviour of polyethersulfone UF membranes made with different PVP. J Membr Sci 211(1):1–11

Maximous N, Nakhla G, Wan W, Wong K (2010) Performance of a novel ZrO_2/PES membrane for wastewater filtration. J Membr Sci 352(1–2):222–230

Moghimifar V, Raisi A, Aroujalian A (2014) Surface modification of polyethersulfone ultrafiltration membranes by corona plasma-assisted coating TiO_2 nanoparticles. J Membr Sci 461:69–80

Mulder M (1984) Basic principles of membrane technology, 1st edn. Kluwer Academic Publishers, Dordrecht

Mulder M (1997) Basic principles of membrane technology, 2nd edn. Kluwer Academic Publishers, Dordrecht

Nair PK, Cardoso J, Gomez Daza O, Nair MTS (2001) Polyethersulfone foils as stable transparent substrates for conductive copper sulfide thin film coatings. Thin Solid Films 401(1–2):243–250

Nunes S, Peinemann K (2001) Membrane technology. Wiley, Weinheim

Nunes S, Peinemann K (2006) Membrane technology. Wiley, Weinheim

Nunes S, Peinemann K (2010) Membranes for water treatment, 4th edn. Wiley, Weinheim

Oh SJ, Kim N, Lee YT (2009) Preparation and characterization of PVDF/TiO$_2$ organic-inorganic composite membranes for fouling resistance improvement. J Membr Sci 345(1–2):13–20

Peyravi M, Rahimpour A, Jahanshahi M (2012) Thin film composite membranes with modified polysulfone supports for organic solvent nanofiltration. J Membr Sci 423–424:225–237

Rahimpour A (2011) Preparation and modification of nano-porous polyimide (PI) membranes by UV photo-grafting process: ultrafiltration and nanofiltration performance. Korean J Chem Eng 28(1):261–266

Rahimpour A, Jahanshahi M, Peyravi M, Khalili S (2011) Interlaboratory study of highly permeable thin film composite polyamide nanofiltration membrane. Polym Adv Technol 23:884–893

Rahimpour A, Madaeni SS (2007) Polyethersulfone (PES)/cellulose acetate phthalate (CAP) blend ultrafiltration membranes: preparation, morphology, performance and antifouling properties. J Membr Sci 305(1–2):299–312

Rahimpour A, Madaeni SS (2010) Improvement of performance and surface properties of nano-porous polyethersulfone (PES) membrane using hydrophilic monomers as additives in the casting solution. J Membr Sci 360(1–2):371–379

Rahimpour A, Madaeni SS, Mansourpanah Y (2007a) The effect of anionic, non-ionic and cationic surfactants on morphology and performance of polyethersulfone ultrafiltration membranes for milk concentration. J Membr Sci 296(1–2):110–121

Rahimpour A, Madaeni SS, Mansourpanah Y (2007b) High performance polyethersulfone UF membrane for manufacturing spiral wound module: preparation, morphology, performance, and chemical cleaning. Polym Adv Technol 18(5):403–410

Rahimpour A, Madaeni SS, Mansourpanah Y (2010a) Fabrication of polyethersulfone (PES) membranes with nano-porous surface using potassium perchlorate (KClO$_4$) as an additive in the casting solution. Desalination 258(1–3):79–86

Rahimpour A, Madaeni SS, Mansourpanah Y (2010b) Nano-porous polyethersulfone (PES) membranes modified by acrylic acid (AA) and 2-hydroxyethylmethacrylate (HEMA) as additives in the gelation media. J Membr Sci 364(1–2):380–388

Rahimpour A, Madaeni SS, Taheri AH, Mansourpanah Y (2008) Coupling TiO$_2$ nanoparticles with UV irradiation for modification of polyethersulfone ultrafiltration membranes. J Membr Sci 313(1–2):158–169

Rahimpour A, Madaeni SS, Zereshki S, Mansourpanah Y (2009) Preparation and characterization of modified nano-porous PVDF membrane with high antifouling property using UV photo-grafting. Appl Surf Sci 255(16):7455–7461

Razali NF, Mohammad AW, Hilal N, Leo CP, Alam J (2013) Optimisation of polyethersulfone/ polyaniline blended membranes using response surface methodology approach. Desalination 311:182–191

Razmjou A, Mansouri J, Chen V (2011a) The effects of mechanical and chemical modification of TiO$_2$ nanoparticles on the surface chemistry, structure and fouling performance of PES ultrafiltration membranes. J Membr Sci 378(1–2):73–84

Razmjou A, Mansouri J, Chen V, Lim M, Amal R (2011b) Titania nanocomposite polyether-sulfone ultrafiltration membranes fabricated using a low temperature hydrothermal coating process. J Membr Sci 380(1–2):98–113

Samtleben W, Dengler C, Reinhardt B, Nothdurft A, Lemke HD (2003) Comparison of the new polyethersulfone high-flux membrane DIAPES® HF800 with conventional high-flux membranes during on-line haemodiafiltration. Nephrol Dial Transplant 18(11):2382–2386

Sawada I, Fachrul R, Ito T, Ohmukai Y, Maruyama T, Matsuyama H (2012) Development of a hydrophilic polymer membrane containing silver nanoparticles with both organic antifouling and antibacterial properties. J Membr Sci 387–388(1):1–6

Scott JA, Neilson DJ, Liu W, Boon PN (1998) A dual function membrane bioreactor system for enhanced aerobic remediation of high-strength industrial waste. Water Sci Technol 38(4–5):413–420

Seidel A, Elimelech M (2002) Coupling between chemical and physical interactions in natural organic matter (NOM) fouling of nanofiltration membranes: implications for fouling control. J Membr Sci 203(1–2):245–255

Seman MA, Khayet M, Hilal N (2012) Development of anti-fouling properties and performance of nanofiltration membranes by interfacial polymerization and photografting In: Hilal N, Khayet M, Wright B (eds) Membrane modification technology and applications. Boca Raton, Taylor & Francis Group: CRC press, pp 119–153

Shi Q, Su Y, Zhao W, Li C, Hu Y, Jiang Z, Zhu S (2008) Zwitterionic polyethersulfone ultrafiltration membrane with superior antifouling property. J Membr Sci 319(1–2):271–278

Shin SJ, Kim JP, Kim HJ, Jeon JH, Min BR (2005) Preparation and characterization of polyethersulfone microfiltration membranes by a 2-methoxyethanol additive. Desalination 186 (1–3):1–10

Song YQ, Sheng J, Wei M, Yuan XB (2000) Surface modification of polysulfone membranes by lowtemperature plasma-graft poly(ethylene glycol) onto polysulfone membranes. J Appl Polym Sci 78(5):979–985

Strathmann, H (1989) Economical evaluation of membrane technology In: Cecille L, Toussaint JC (eds) Future industrial prospects of membrane processes. Elsevier, London & New York, pp 41–55

Strathmann H (2000) Introduction to membrane science and technology. Wiley-VCH, Weinheim

Strathmann H (2011) Introduction to membrane science and technology. Wiley-VCH, Weinheim

Strathmann H, Giorno L, Drioli E (2006) An introduction to membrane science and technology. In: Consiglio Nazionale delle Ricerche, Rome

Su Y, Mu C, Li C, Jiang Z (2009a) Antifouling property of a weak polyelectrolyte membrane based on poly(acrylonitrile) during protein ultrafiltration. Ind Eng Chem Res 48(6):3136–3141

Su YL, Cheng W, Li C, Jiang Z (2009b) Preparation of antifouling ultrafiltration membranes with poly(ethylene glycol)-graft-polyacrylonitrile copolymers. J Membr Sci 329(1–2):246–252

Tang B, Huo Z, Wu P (2008) Study on a novel polyester composite nanofiltration membrane by interfacial polymerization of triethanolamine (TEOA) and trimesoyl chloride (TMC). I. Preparation, characterization and nanofiltration properties test of membrane. J Membr Sci 320(1–2):198–205

Tsuru T, Izumi S, Yoshioka T, Asaeda M (2000a) Temperature effect on transport performance by inorganic nanofiltration membranes. AIChE J 46(3):565–574

Tsuru T, Sudou T, Kawahara SI, Yoshioka T, Asaeda M (2000b) Permeation of liquids through inorganic nanofiltration membranes. J Colloid Interface Sci 228(2):292–296

Tullis RH, Duffin RP, Zech M, Ambrus JL Jr (2002) Affinity hemodialysis for antiviral therapy. I. Removal of HIV-1 from cell culture supernatants, plasma, and blood. Ther Apheresis 6(3):213–220

Ulbricht M (2006) Advanced functional polymer membranes. Polymer 47(7):2217–2262

Ulbricht M, Schuster O, Ansorge W, Ruetering M, Steiger P (2007) Influence of the strongly anisotropic cross-section morphology of a novel polyethersulfone microfiltration membrane on filtration performance. Sep Purif Technol 57(1):63–73

Unger RE, Peters K, Huang Q, Funk A, Paul D, Kirkpatrick CJ (2005) Vascularization and gene regulation of human endothelial cells growing on porous polyethersulfone (PES) hollow fiber membranes. Biomaterials 26(17):3461–3469. doi:10.1016/j.biomaterials.2004.09.047

Van der Bruggen B, Braeken L, Vandecasteele C (2002a) Evaluation of parameters describing flux decline in nanofiltration of aqueous solutions containing organic compounds. Desalination 147 (1–3):281–288

Van der Bruggen B, Braeken L, Vandecasteele C (2002b) Flux decline in nanofiltration due to adsorption of organic compounds. Sep Purif Technol 29(1):23–31

van Veen HM, Bracht M, Hamoen E, Alderliesten PT (1996) Chapter 14 Feasibility of the application of porous inorganic gas separation membranes in some large-scale chemical processes. Membr Sci Technol 4

Wang Y, Su Y, Sun Q, Ma X, Ma X, Jiang Z (2006a) Improved permeation performance of Pluronic F127-polyethersulfone blend ultrafiltration membranes. J Membr Sci 282(1–2):44–51

Wang YQ, Su YL, Ma XL, Sun Q, Jiang ZY (2006b) Pluronic polymers and polyethersulfone blend membranes with improved fouling-resistant ability and ultrafiltration performance. J Membr Sci 283(1–2):440–447

Wang YQ, Wang T, Su YL, Peng FB, Wu H, Jiang ZY (2006c) Protein-adsorption-resistance and permeation property of polyethersulfone and soybean phosphatidylcholine blend ultrafiltration membranes. J Membr Sci 270(1–2):108–114

Werner C, Jacobasch HJ, Reichelt G (1995) Surface characterization of hemodialysis membranes based on streaming potential measurements. J Biomater Sci Polym Ed 7(1):61–76

Wu H (2012) Improving the anti-fouling and fouling release of PVDF UF membrane by chemically modified SiO2 nanoparticles. New South Wales, Sydney, Australia

Xu ZL, Qusay FA (2004) Polyethersulfone (PES) hollow fiber ultrafiltration membranes prepared by PES/non-solvent/NMP solution. J Membr Sci 233(1–2):101–111

Yamamura H, Kimura K, Watanabe Y (2007a) Mechanism involved in the evolution of physically irreversible fouling in microfiltration and ultrafiltration membranes used for drinking water treatment. Environ Sci Technol 41(19):6789–6794

Yamamura H, Okimoto K, Kimura K, Watanabe Y (2007b) Influence of calcium on the evolution of irreversible fouling in microfiltration/ultrafiltration membranes. J Water Supply Res Technol —AQUA 56(6–7):425–434

Yang Y, Zhang H, Wang P, Zheng Q, Li J (2007) The influence of nano-sized TiO2 fillers on the morphologies and properties of PSF UF membrane. J Membr Sci 288(1–2):231–238

Yu LY, Shen HM, Xu ZL (2009a) PVDF-TiO2 composite hollow fiber ultrafiltration membranes prepared by TiO2 sol-gel method and blending method. J Appl Polym Sci 113(3):1763–1772

Yu LY, Xu ZL, Shen HM, Yang H (2009b) Preparation and characterization of PVDF-SiO2 composite hollow fiber UF membrane by sol-gel method. J Membr Sci 337(1–2):257–265

Yu Y, Seo S, Kim IC, Lee S (2011) Nanoporous polyethersulfone (PES) membrane with enhanced flux applied in forward osmosis process. J Membr Sci 375(1–2):63–68. doi:10.1016/j.memsci.2011.02.019

Zhao C, Liu T, Lu Z, Cheng L, Huang J (2001) An evaluation of a polyethersulfone hollow fiber plasma separator by animal experiment. Artif Organs 25:3–60

Zhao C, Liu X, Rikimaru S, Nomizu M, Nishi N (2003a) Surface characterization of polysulfone membranes modified by DNA immobilization. J Membr Sci 214(2):179–189

Zhao C, Xue J, Ran F, Sun S (2013) Modification of polyethersulfone membranes—A review of methods. Prog Mater Sci 58(1):76–150

Zhao Q, Qian J, An Q, Zhu Z, Zhang P, Bai Y (2008) Studies on pervaporation characteristics of polyacrylonitrile-b-poly(ethylene glycol)-b-polyacrylonitrile block copolymer membrane for dehydration of aqueous acetone solutions. J Membr Sci 311(1–2):284–293

Zhao ZP, Wang Z, Wang SC (2003b) Formation, charged characteristic and BSA adsorption behavior of carboxymethyl chitosan/PES composite MF membrane. J Membr Sci 217(1–2):151–158

Zhenxin Z, Matsuura T (1991) Discussions on the formation mechanism of surface pores in reverse osmosis, ultrafiltration, and microfiltration membranes prepared by phase inversion process. J Colloid Interface Sci 147(2):307–315

Zhu J, Guo N, Zhang Y, Yu L, Liu J (2014) Preparation and characterization of negatively charged PES nanofiltration membrane by blending with halloysite nanotubes grafted with poly (sodium 4-styrenesulfonate) via surface-initiated ATRP. J Membr Sci 465:91–99

Zoppi RA, Soares CGA (2002) Hybrids of poly(ethylene oxide-b-amide-6) and ZrO_2 sol-gel: Preparation, characterization, and application in processes of membranes separation. Adv Polym Technol 21(1):2–16

Chapter 3
Fouling in Membrane Bioreactors

Membrane fouling is a stubborn problem in all membrane filtration processes, in particular membrane bioreactors because it leads to higher operating pressure, more frequent chemical cleaning, shortened membrane life and compromised product water quality. This chapter presents an exhaustive overview of membrane fouling in membrane bioreactors. It commences by giving a concise definition of membrane fouling and its diverse implication in the development of membrane bioreactor technology. This chapter highlights the underlying causes of membrane fouling and its effects are also indicated. The types of membrane fouling in membrane bioreactors are elucidated in detail. Thereafter, methods used to control or limit membrane fouling are also outlined in this chapter. To sum up, membrane fouling is highly complex physico-chemical problem.

3.1 Membrane Fouling

Generally, membrane fouling is considered as a bottleneck in membrane filtration processes (Huang et al. 2012b). It is inevitable problem in which the retained particles, colloids, macromolecules and salts are undesirably deposited and accumulated on the membrane surface or in the membrane pores (Houari et al. 2010; Meng et al. 2009; Rana and Matsuura 2010; Kochkodan and Hilal 2015). Membrane fouling is a very common hindrance to the advancement of water treatment membrane technologies, including microfiltration (Xiao et al. 2011); ultrafiltration (Bai and Leow 2002a, b; Le-Clech et al. 2006a, b), nanofiltration (Mo et al. 2012; Simon et al. 2013) and osmosis processes (Li et al. 2007; Phuntsho et al. 2012). Therefore, membrane fouling decreases the permeate flux significantly, affects the quantity and quality of products, raises operating costs, and eventually shortens membrane lifespan (Wang et al. 2011a, b; Porcelli and Judd 2010).

© Springer Nature Singapore Pte Ltd. 2017
B. Ladewig and M.N.Z. Al-Shaeli, *Fundamentals of Membrane Bioreactors*,
Springer Transactions in Civil and Environmental Engineering,
DOI 10.1007/978-981-10-2014-8_3

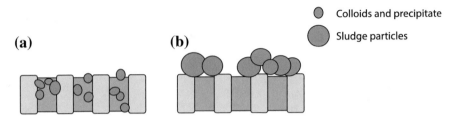

Fig. 3.1 Membrane fouling process in membrane bioreactors, via **a** pore blocking and **b** cake layer formation. Adapted from Meng et al. (2009)

With respect to MBR, membrane fouling is one of the most stubborn problems, hindering its widespread practical applications and also reduces its performance (Hilal et al. 2005; Huang et al. 2010; Kang et al. 2003; Kimura et al. 2005; Le-Clech et al. 2006a, b; Meng et al. 2009; Zularisam et al. 2006; Miura et al. 2007; Wang et al. 2011a, b).

Membrane fouling can be ascribed to both pore clogging and sludge cake deposition which are regarded as the main fouling components as shown schematically in Fig. 3.1 (Hilal et al. 2005; Lee et al. 2001a), whilst other adsorption of solutes on membranes, deposition of particle within the membrane pores and alterations to the cake layer affect membrane fouling via the modification of either or both components (Bai and Leow 2002a, b; Ma et al. 2001a, b; Wiesner et al. 1992; Ahmed 1997; Wakeman and Williams 2002a). Pore blocking and adsorption in internal pore surfaces occur if the foulants (colloids) are smaller than the membrane pores (i.e. solutes). However, if the foulants (colloids and sludge flocs) are much larger than the membrane pores, they tend to form cake layer on the surface of membrane. In fact, pore blocking increases the membrane resistance whereas cake layer can create an additional layer of resistance to permeation flow (Bai and Leow 2002a; Wiesner et al. 1992).

During the past few decades, there have been numerous research studies conducted to grasp the complex mechanisms of membrane fouling and strategies that can be implemented to keep the filterability of membranes which in turn controls fouling (Tardieu et al. 1998; Xiao et al. 2011; Ang et al. 2011; Chon et al. 2013; Lee et al. 2004; Asatekin et al. 2007; Hilal et al. 2005; Hong et al. 2002; Le-Clech et al. 2006a, b, c). However, there are still research gaps on how to counteract fouling successfully (Wu and Huang 2010a; Wang et al. 2011; Tiraferri et al. 2012; Diagne et al. 2012; Cui and Choo 2013).

Fouling has been a problematic phenomenon the industry has been battling for a while, and extensive research must be conducted on this issue. This can broaden our understanding of fouling which can result in finding an effective and efficient technique to control and even minimise it. Once a feasible solution for membrane fouling is implemented, membrane bioreactors will become a more promising technology for a wider range of wastewater treatment.

3.1.1 Causes of Fouling

Huang et al. (2012b), Le-Clech et al. (2006b), Rana and Matsuura (2010) and Zhou et al. (2014) state that fouling is caused by the interaction between the foulants which may be particulate, colloidal particles or matters or biomacromolecules in separation solutions and the membrane surface which includes: organic, inorganic and biological substances in numerous forms. The foulants interact physically and chemically with the membrane surface, however chemically it degrades the membrane material. Consequently, non-specific adhesion of microorganisms and biomacromolecules occurs on the membrane surface, resulting in blocked or decreased greatly membrane pores and then a significant decrease in permeation flux or separation efficiency.

Previous studies have indicated that the factors which affect membrane fouling in membrane bioreactors are: the type of wastewater (Li and Yang 2007), sludge age (Chang and Lee 1998), sludge loading rate (Chang et al. 2002), permeate flux (Tradieu et al. 1998; Fan et al. 2000), aeration intensity (Bouhabila et al. 1998; Howell et al. 2004), mixed liquor suspended solid concentration (MLSS) (Hong et al. 2002; Le-Clech et al. 2003; Yamamoto et al. 1989), mechanical stress (Zeng 2007), solid retention time (SRT) (Shin and Kang 2003; Lee et al. 2003), food to microorganism ratio (F/M), and hydraulic retention time (HRT) (Rosenberger and Kraume 2002; Rosenberger et al. 2002a, b). Table 3.1 gives the relationship between these factors and membrane fouling on the basis of recent research studies.

In addition to the factors above, the properties of mixed liquor have also been thought to impact membrane fouling in MBRs. These properties include soluble compounds (Wisniewski and Grasmick 1998), soluble microbial products (SMP) (Huang et al. 2000; Lesjean et al. 2005; Liu et al. 2005), extracellular polymeric substances (EPS) (Chang and Lee 1998), particle size distribution(Cicek et al. 1999) and viscosity of mixed liquor (Ueda et al. 1996). Furthermore, the research studies also illustrate that carbohydrate/protein has a solid relationship with the evolution of membrane fouling in membrane bioreactor systems. However, a clear relationship between carbohydrate/protein and membrane fouling in MBRs was not apparently found in recent research studies in which pilot-scale experiments were implemented by Kimura et al. (2005). Other investigators, for example (Evenblij and van der Graaf 2004; Nuengjamnong et al. 2005; Drews et al. 2006a; Geng and Hall 2007; Nagaoka et al. 1996; Ng et al. 2006) have concluded that EPS and SMP are currently regarded as the major foulants of membranes in MBRs. The deposition of EPS and SMP towards membranes can clog membrane pores and form a fouling layer on the membrane surface, gradually increasing the filtration resistance (Tansel et al. 2006; Brindle and Stephenson 1996; Chang and Lee 1998; Cho and Fane 2002; Drews et al. 2007; Rosenberger and Kraume 2002). Zhang et al. (2011a) vividly outlines the factors impacting membrane fouling in membrane bioreactors (MBRs). The authors state that the primary cause of fouling originates from three main parameters including membrane properties, sludge characteristics and operating parameters; this can be seen in Fig. 3.2. The operating parameters

Table 3.1 Summary of the relationship between various fouling factors and membrane fouling (Meng et al. 2009)

Sludge condition	Effect on membrane fouling	Reference
MLSS	– High MLSS leads to normalised lower permeability – High MLSS leads to higher fouling potential – High MLSS leads to lower cake resistance, lower specific cake resistance	Trussell et al. (2007) Psoch and Schiewer (2006) Chang and Kim (2005)
Viscosity	– High viscosity leads to low membrane permeability – High MLSS/Viscosity leads to membrane permeability – High viscosity leads to high membrane resistance	Li et al. (2007) Trussell et al. (2007) Chae et al. (2006)
F/M	– High F/M ratio leads to high fouling rates – MLSS (2–3 g/L): high F/M leads to high irremovable fouling – MLSS (8–12 g/L): high F/M leads to higher removable fouling – High F/M leads to high protein in foulants	Trussell et al. (2006) Watanabe et al. (2006) Kimura et al. (2005)
EPS	– Increased polysaccharide concentration leads to increased fouling rate – Bound EPS influences on specific cake resistance – Increased polysaccharide concentration leads to increased fouling rate – Increased bound EPS leads to membrane resistance – The loosely bound EPS contributes to most of the filtration resistance of the whole sludge	Drews et al. (2006b) Cho et al. (2005b) Lesjean et al. (2005) Chae et al. (2006) Ramesh et al. (2007)
SMP	– SMP is more important than MLSS – Colloidal TOC relates with permeate flux – Filtration resistance is determined by SMP – SMP is probably responsible for fouling – Polysaccharide is a possible indicator of fouling – Low SMP leads to low fouling index – Fouling rates correlate with SMP	Zhang et al. (2006b) Fan et al. (2006) Jeong et al. (2007) Sperandio et al. (2005) Le-Clech et al. (2005b) Jang et al. (2006) Trussell et al. (2006)
Filamentous bacteria	– High filamentous bacteria concentration leads to high sludge viscosity – Bulking sludge could cause a severe fouling – Low filamentous bacteria concentration leads to cake resistance	Meng et al. (2007) Sun et al. (2007) Kim and Jang (2006)
Operating condition SRT	– Decreased SRT from 100 to 20 d leads to high TMP – Decreased SRT from 30 to 10 d leads to high fouling – Decreased SRTs leads to high fouling potentials of SMP – Decreased SRT from 5 to 3 d leads to high fouling	Ahmad (2007) Zhang et al. (2006b) Liang et al. (2007) Cho et al. (2005a)

(continued)

Table 3.1   (continued)

Sludge condition	Effect on membrane fouling	Reference
HRT	– High HRT leads to high membrane fouling – Lower HRT leads to high membrane fouling – Lower HRT leads to high membrane fouling	Meng et al. (2007a) Chae et al. (2006) Cho et al. (2005)
Aeration	– High aeration intensity leads to (high) permeability – Air sparging improves membrane flux – Larger bubbles for fouling control are preferable – Air backwashing for fouling control is preferable – Bubble-induced shear reduces fouling significantly – Air scouring can prolong membrane operation	Trussell et al. (2007) Psoch and Schiewer (2006) Phattaranawik et al. (2007) Chae et al. (2006) Wicaksana et al. (2006) Sofia et al. (2004)
Permeate flux	– Subcritical flux mitigates irremovable fouling – Subcritical flux mitigates fouling	Lebegue et al. (2008) Guo et al. (2007)

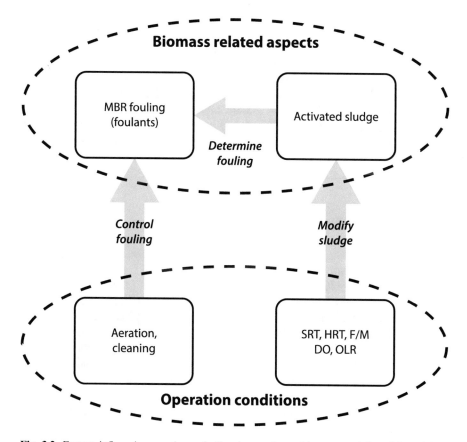

Fig. 3.2 Factors influencing membrane fouling in membrane bioreactors. Adapted from Meng et al. (2009)

[e.g. dissolved oxygen (DO), solid retention time (SRT), hydraulic retention time (HRT) and food to microorganism ratio (F/M)] have directly no impact on membrane fouling but they determine the sludge characteristics. The optimisation of the operating parameters can modify activated sludge. In terms of sludge characteristics, their effects on membrane fouling are very complicated. Therefore, numerous investigators recently add coagulant or adsorbent into membrane bioreactor systems artificially to modify activated sludge. Their results were very effective in reducing membrane fouling (Ji et al. 2008, 2010; Hwang et al. 2007; Koseoglu et al. 2008; Teychene et al. 2011). Although all these research studies have been performed on membrane fouling, they are still insufficient and many questions remain unanswered to date.

3.1.2 Effects of Fouling

Membrane fouling has a number of effects. First of all, it reduces the membrane permeate flux either permanently or temporarily. If the fouling is temporary, then the initial flux can usually be recovered by cleaning the membrane or by applying backpressures to the temporarily fouled membrane. Membranes cannot be restored when they are permanently fouled (Rana and Matsuura 2010). Second, it can significantly reduce membrane performance, reduce separation efficiency, increase maintenance and operating costs and rapid increase in transmembrane pressure (TMP), shorten membrane lifespan, lead to more membrane cleaning or replacement (HT et al. 2010; Yang et al. 2006b; Mansouri et al. 2010; Le-Clech et al. 2006b; Shirazi et al. 2010; Wu and Huang 2010b; Wu et al. 2010; Zhang et al. 2011b). It should be noted that although flux decrease is also associated with the phenomenon of concentration polarisation, this is not considered as fouling because it disappears when the filtration process is stopped.

3.1.2.1 Models of Membrane Fouling

Membrane fouling can be mainly described either by external fouling or internal fouling. Internal fouling occurs when particles deposit inside the membrane structure, leading to blocked pores partially or completely, reducing the effective pore diameter and even reducing flow through the membrane (Baker 2012; Ma et al. 2007). The standard blocking model (SBM), the complete blocking model (CBM) and the intermediate blocking model (IBM) are three models used to describe internal fouling. By contrast, at the early stage of filtration, external fouling refers to surface fouling and occurs thoroughly on the membrane surface when the colloids accumulate; the accumulation of aggregated colloids may lead to blocking pores and finally form a cake layer/film on the membrane surface. External fouling can be removed a certain extent via cleaning procedure. The cake filtration model (CFM) is usually used to describe external fouling.

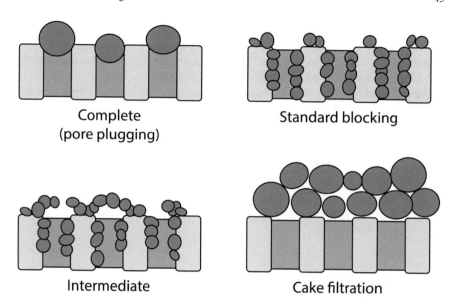

Fig. 3.3 Illustration of several membrane fouling mechanisms. Adapted from Leiknes (2012)

Furthermore, during membrane filtration there have been four fouling models employed to describe the mechanisms of membrane fouling as shown schematically in Fig. 3.3 (Oh et al. 2009).

1. Cake filtration—A uniform cake layer formed upon the entire membrane surface as a consequence of the deposition and accumulation of particles with a larger diameter than the membrane pore size. Cake fouling is generally reversible by water flushing or backwashing (Baker 2012).
2. Complete pore blocking or plugging—this can be caused by occlusion of pores with particles. Superimposition is impossible.
3. Intermediate pore blocking—Similar to complete pore blockage, though the particles have the capability to deposit on the top of other deposited particles. This can only occur under these conditions, superimposition is possible.
4. Standard blocking—Particles with a smaller pore size enter the pores and deposit on the internal pore surfaces with their whole length, causing the narrowing of the pore size.

These fouling models supply a visible picture of the relative position of particles towards membranes. Additionally, they provide a mathematical model of the flux filtration behaviours under constant pressure filtration mode, which can be characterised by a significant plummet in the flux at the beginning of the process, followed by decreasing curve slope until the steady state is achieved. Moreover, developments and extensions make these fouling models applicable for constant flux filtration and spread out some combined models.

3.1.2.2 Types of Fouling

Generally, according to the chemical nature of foulants, membrane process and the types of foulants and their interaction with membrane surface, several types of membrane fouling have been identified in membrane bioreactors (Kochkodan and Hilal 2015; Pan et al. 2010; Flemming 1997; Flemming et al. 1997; Kimura et al. 2004) as shown below:

1. Removable and irremovable fouling.
2. Organic fouling or scaling.
3. Inorganic fouling.
4. Biofouling.
5. Reversible and irreversible fouling.
6. Colloidal fouling.

Removable and irremovable fouling

The control and investigation of both irremovable membrane and removable fouling in MBRs is of significant importance for long-term membrane operation and sustainable operations of MBRs. During initial filtration, solutes, colloids and microbial cells pass through the membrane and precipitate progressively inside the membrane pores. Whilst for long-term operation of MBRs, the deposited cells aggregate in MBRs and form EPS, which is considered as one of the major foulants of membrane in MBRs. EPS is a complex mixture of carbohydrates, proteins, humic compounds, uronic acids and DNA. EPS form a strongly attached fouling layer on the membrane surface and clog the membrane pores partially or completely. Meantime, some inorganic substances might stick progressively on the membrane surface or in the membrane pores (Meng et al. 2009).

Irremovable fouling can be caused by pore blocking and strongly attached foulants during filtration, whilst removal fouling is caused by foulants that loosely attach and is associated by the cake layer formation above the membrane surface. Therefore, to counteract removable and irremovable fouling propensities, many researchers stated that operation below the critical flux is the best approach to control membrane fouling, particularly removable and irreversible fouling within a specified filtration system. For example, (Field et al. 1995) introduced the concept of critical flux. This flux is called subcritical flux or non-fouling operation if the operation is below the critical flux and is expected to cause less irremovable fouling. For a short-term membrane operation, when the permeate flux or membrane permeation is less than the critical flux filtration conditions, the transmembrane pressure (TMP) remains constant and fouling was removable. By contrast, when the permeate flux exceeds the critical flux, the TMP increases and is expected to increase significantly. Indeed, for long-term membrane operation, irremovable fouling can appear if the process is operated below the critical flux. Ognier et al. (2004) stated that with the choice of subcritical flux filtration conditions, a slight membrane fouling was still seen to develop, which proved to be hydraulically irremovable after long-term operation. The critical flux value relies on operating conditions (i.e. aeration intensity, temperature), membrane characteristics and sludge characteristics. The principle of critical flux has gained popularity in the study of MBR fouling (Guglielmi et al. 2007a, b; Lebegue et al. 2008; Wang et al. 2008). However, most of the research studies on the determination of critical flux are based on ex situ measuring which cannot provide reliable information about MBRs fouling. Therefore, in recent years, De La Torre et al. (2008) developed an in situ measuring method in which more reliable information can provide about critical flux in comparison to ex situ measuring method. This method is superior to the ex situ measuring method. Another research study was conducted by Huyskens et al. (2008), using an online measuring method. In this study, the removable and irremovable fouling intensities of MBR were evaluated in a reproducible way. This method seems to be more practical approach to evaluate fouling in MBRs. These research studies state that it is possible to develop online method or in situ method to determine removable/irremovable fouling or determine critical flux. It is also of great interest to develop an apparatus or a unified measuring method to verify membrane propensity.

As mentioned previously, irremovable fouling plays an important role in the long-term MBR operation. Therefore, chemical cleaning is necessary to maintain MBR operation. But, chemical cleaning for the sake of eliminating irremovable membrane fouling should be limited to a minimum because repeated chemical cleaning may shorten membrane lifespan, degrade the membrane and the disposal of spent chemical agents causes environmental issues (the chemical agents are not friendly to environment as it leads to pollution of water) (Yamamura et al. 2007a, b, c). On the other hand, removable fouling can be eliminated by the execution of physical cleaning (e.g. backwashing/backflushing/relaxation) (Meng et al. 2009). Figure 3.4 depicts the process of formation of removable and irremovable fouling and their elimination in membrane bioreactors.

Organic fouling

Schafer et al. (2005) outlines that organic fouling is an irreversible process caused by the adsorption/retention of dissolved organic materials. Whilst, Meng et al. (2009) define organic fouling as the deposition of (soluble microbial product) SMP or NOM (natural organic matter) which is composed of proteins, polysaccharides and humic substances on the surface of membrane bioreactors. SMP or NOM can be easily retained onto the membrane surface due to the small size and the permeate flow as the back transport to the bulk phase due to lift forces is very weak compared with large particles (e.g. colloids and sludge flocs). For NOM, humic substances have been considered as major foulants which can cause severe irreversible foulants in pressure-driven membrane processes through membrane adsorption and pore blocking (Escobar 2005). This results in a flux decline. Also polysaccharides (with large particle size) are of high importance for membrane fouling when pressure-driven membrane (UF) is used for water recycling (Laabs et al. 2006).

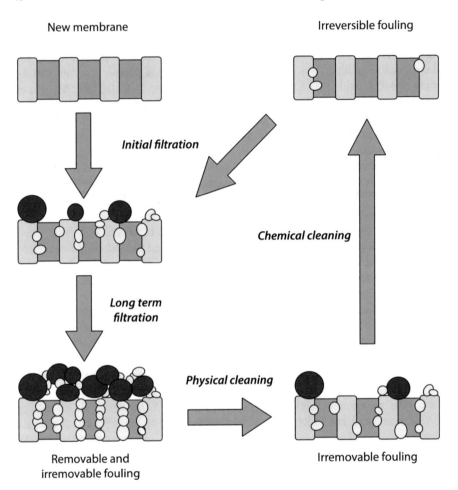

New membrane

Irreversible fouling

Initial filtration

Chemical cleaning

Long term filtration

Physical cleaning

Removable and irremovable fouling

Irremovable fouling

Fig. 3.4 Schematic illustration of the formation and removal of removable and irremovable fouling in membrane bioreactors. Adapted from Meng et al. (2009), Pollice et al. (2005)

It is difficult to recognise the mechanism attributed with organic fouling due to the chemical properties of the molecule, the interactions of chemicals properties with the membrane materials and strong effects of organic types. Nevertheless, colloidal fouling and organic fouling usually intertwine for some reasons. First, the colloids adsorb organic matter in the natural environment and a negatively charged surface to stabilise them. Second, some organics with the sizes ranging from 1 nm to 1 μm can be referred to colloids, such as natural organic matter (NOM). Up until now, the only **two vital mechanisms** that can explain organic fouling are gel layer formation and initial adsorption.

Gel layer is formed as a result of an excess solubility of organic molecules and the appearance of organic flocculation. Generally, it is attributed with the concentration polarisation when the rise in concentration reaches the top layer of the membrane surface. Many investigators identify Natural Organic Matter (NOM) as the major foulant associated with organic fouling (Al-Amoudi and Farooque 2005; Rana and Matsuura 2010; Roudman and Digiano 2000). Kim et al. (2007) characterised the formation of NOM on flat sheet membrane surface using dead-end cell filtration. The results of Scanning Emission Microscope (SEM) indicate that NOM gets accumulated and a gel layer is formed on the membrane surface though it does not breach the membrane body. In contrast to other research studies, NOM can reduce the effective pore size by adsorbing on the pore wall. Rana and Matsuura (2010) state that NOM could be controlled by electric double layer repulsion and permeation hindrance. It has been demonstrated that the hydrophobic fraction of NOM was the major factor in causing the flux decline. This is because of the strong adsorption of hydrophobic NOM on the membrane surface. The hydrophilic fraction of NOM had a relatively small effect on the membrane fouling (Nilson and DiGiano 1996). Recently, (Metzger et al. 2007) have performed a comprehensive research study to characterise the deposited biopolymers in MBRs. After membrane filtration, the fouling layers were divided into three layers (upper layer, intermediate layer and lower layer) using rinsing (washing), backwashing and chemical cleaning. The results showed that the upper fouling layer was composed of a porous, loosely bound cake layer with the same composition to sludge flocs (biomass flocs). The intermediate fouling layer equally consisted of soluble microbial product (SMP) and biomass aggregates, resulted in a high concentration of polysaccharides. Whilst, the lower layer was predominated by SMP and represented the irremovable fouling fraction, which had a relative higher concentration of strongly bound proteins. This study showed the spatial distribution of SMP on the surface of membrane.

Whilst adsorption occurs when organic matter and membrane interact physico-chemically. Combe et al. (1999) confirmed that the organic components directly adsorb to the membrane surface or inside the pores in a form of a thin layer, using humic acid as the foulant model. Hence, the characteristics of the membrane surface such as hydrophobicity, surface charges and pore size change. The aforementioned characteristics result in fouling to some extent. It should be noticed that after the initial adsorption of the organics, the newly formed thin surface might have completely distinct characteristics compared with the original membrane surface, which mainly depends on the solutes. Under this consideration, gel layer formation seems to be more responsible for the organic fouling.

In order to figure out the detailed information on the deposited biopolymers (i.e. Protein, humic substances, polysaccharides), the identification of these substances is indispensible part. Fourier transform infrared spectroscopy (FTIR), excitation–emission matrix fluorescence spectroscopy (three dimension tool) EEM and high performance size exclusion chromatography (HOP-SEC) are three analytical tools used to investigate organic fouling. The major components of NOM foulants were identified as protein using FTIR (Lee et al. 2004) whilst EEM were usually used to characterise protein-like or humic-like substances in membrane foulants.

Inorganic fouling or scaling

Fouling in membrane bioreactors is mainly dominated by organic fouling and biofouling, whilst inorganic fouling is the weakest threat. All of them occur simultaneously during the membrane filtration of activated sludge (Meng et al. 2009). Despite numerous research studies focused on membrane fouling, not much light has been shed on inorganic fouling in MBRs. Speth et al. (1998) define inorganic fouling as the agglomeration of materials on the surface of a membrane, or in membrane pores. Van De Lisdonk et al. (2000), Kochkodan and Hilal (2015) contend that inorganic fouling is the result of high concentration of one or more inorganic salts in raw water beyond the inorganic salts limited solubility and their ultimate precipitation on the membranes. Conspicuously, this can be seen vividly in the region near the surface of nanofiltration membranes NF and also near the RO membrane surface where the concentration of dissolved salts is 4–10 times higher than the bulk feed water. This is due to concentration polarisation (CP), resulting in the precipitation of constituent ions (e.g. $BaSO_4$ and $CaCO_3$) on membrane surface (Van De Lisdonk et al. 2000; Wu 2012). It is documented that inorganic substances are a contributor factor to less than 15 % of membrane foulants in pilot nanofiltration plant for the treatment of surface water. A severe fouling is expected when the presence of calcium and alginate is due to the complexation between them.

Scaling refers to the formation of inverse solubility salts, such as $CaSO_4.xH_2O$, $CaCO_3$, SiO_2, $Mg(OH)_2$ and $Ca_3(PO_4)_2$ deposits. These salts are responsible for inorganic fouling on the membrane surface and their solubility products are illustrated below in Table 3.2 (Lee et al. 1999; Van De Lisdonk et al. 2000; Lin et al. 2005; Shirazi et al. 2006). According to (Hasson et al. 2001), the compounds with the greatest scaling in NF and RO are $CaCO_3$ and $CaCO_4.2H_2O$, whilst other potential scaling compounds are $BaSO_4$, $SrSO_4$, $Ca_3(PO_4)_2$ and $Fe(OH)_3$. The salt precipitates when the solubility product of the constituent ions exceeds the solubility limit. Table 3.2 illustrates the solubility product of the common inorganic salts that cause scaling on the surface of the membrane.

Other investigators, for example Kang et al. (2002) and Ognier et al. (2002a, b, c) suggest that inorganic fouling may take place more readily on inorganic membranes. Due to the cohesive properties, a cake of inorganic material can generally become irremovable. Recently, Wang et al. (2008) asserted that the cake layer is formed by organic substances and inorganic elements such as Mg, Al, Fe, Ca, Si, etc. When the organic foulants are coupled with the inorganic foulants, precipitation promotes the formation of a cake layer.

Recently, inorganic fouling has been investigated in MBR processes. Ognier et al. (2002a, b, c) state that severe $CaCO_3$ fouling is expected in a pilot MBR with a ceramic ultrafiltration membrane. They prepared synthetic wastewater with hard tap water ($Ca^{+2} = 120$ mg/l; $Mg^{+2} = 8$ mg/l). They concluded that higher alkalinity of the activated sludge (pH = 8–9) leads to pre-precipitation of $CaCO_3$. Lyko et al. (2007) found that metal substances (i.e. Fe^{+3}) were regarded as major contributors to membrane fouling than biopolymers (i.e. protein, polysaccharides). You et al. (2006) point out that the fouling caused by inorganic scaling is not readily removed

Table 3.2 Solubility products of the common inorganic salts that cause scaling on membrane surfaces (Pollice et al. 2005; Xie et al. 2004)

Fouling salt	Solubility product
$CaCO_3$	2.8×10^{-9}
$CaHPO_4$	1×10^{-7}
$CaSO_4$	4.93×10^{-5}
$Ca_3(PO_4)_2$	2.07×10^{-29}
$MgCO_3.3H_2O$	2.38×10^{-6}
$Mg_3(PO_4)_2$	1.04×10^{-24}
$AlPO_4$	9.83×10^{-21}
$Al(OH)_3$	1.3×10^{-33}
$Ca(OH)_2$	5.5×10^{-6}
$CaHPO_4$	1.0×10^{-9}
$Fe(OH)_3$	2.79×10^{-39}
$FePO_4.2H_2O$	9.92×10^{-29}

by chemical cleaning. These research studies demonstrate that inorganic fouling has become paramount in MBRs. The understanding of inorganic fouling however is still vague. Studies conducted on metal ions with limited concentrations submerged in the feed of wastewater may promote inorganic fouling, and this may be an important area of research in the foreseeable future. Even though, the chemical composition of wastewater is directly correlated with precipitants formed.

As noted by (Meng et al. 2009), inorganic fouling can be formed by two paths as shown schematically in Fig. 4.6. The first way is referred as **chemical precipitation** and the second one is **biological precipitation**. In membrane bioreactors, a variety of cations and anions and others are present such as Ca^{2+}, Mg^{2+}, Al^{3+}, Fe^{3+}, CO_3^{2+}, SO_4^{2-}, PO_4^{3-} and OH^-. Concentration polarisation will lead to higher concentration of deposited salts on the surface of membrane, particularly for applications of RO and NF.

Chemical precipitation on the one hand, takes place when the concentration of chemical species exceeds the saturation concentrations owing to concentration polarisation. Furthermore, as biocake or biofilm is naturally elastic, In addition, the fouling layer on the membrane surface can protect the surface layer from shear stress as biocake is naturally elastic, resulting in higher concentration polarisation (CP) and precipitation of inorganic salts (Sheikholaslami 1999; Sheikholeslami 1999). Carbonates are one type of the predominant salts in inorganic fouling. The carbonates of metals such as Ca, Mg and Fe can potentially increase membrane scaling (You et al. 2005).

Biological precipitation, by contrast, is another factor to inorganic fouling. The biopolymers (i.e. Protein) contain negative ions such as COO^-, $CO3^{2-}$, $SO4^{2-}$, $PO4^{3-}$ and OH^-. The metal ions can be readily caught by these negative ions. Acidic functional groups (R–COOH) and calcium (Ca^{+2}), in some cases, can produce complexes and build a dense gel layer (a network of rigid organic matter) that may lead to flux decline (Costa et al. 2006). When the metal ions pass through the membranes, they could be captured by the attached biocake layer on membrane

Fig. 3.5 Schematic illustration of the formation of inorganic fouling in membrane bioreactors, as well as a representation of biological precipitation. Adapted from Meng et al. (2009), Pollice et al. (2005)

Chemical Precipitation

$$M^{n+} + nOH^- \rightarrow M(OH)_n$$
$$M^{n+} + CO_3^{2-} \rightarrow MCO_3$$
$$M^{n+} + OH^- + CO_2 \rightarrow MCO_3$$
$$M^{n+} + SO_4^{2-} \rightarrow MSO_4$$

Biological Precipitation

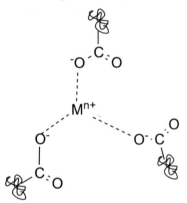

through charge neutralisation and complexation and therefore speed up membrane fouling. Connection between the deposited cells and biopolymers further enhanced the compactness of the fouling layer and then form a dense cake layer on membranes (Hong et al. 1997). The synergistic interaction between different types of fouling (biofouling, organic fouling and inorganic fouling) could result in more foulants deposited on the surface of membrane. Figure 3.5 illustrates chemical and biological precipitation.

Bear in mind, inorganic fouling is a complex phenomenon in membrane bioreactors. Therefore, different methods have been performed to avert inorganic fouling. One of these methods is pretreatment of feed water or use of chemical cleaning. Chemical cleaning is very straightforward method and simpler than physical one in removal of inorganic precipitation as inorganic fouling can result in severe irremovable fouling. Meng et al. (2009) propose using Ethylenediaminetetraacetic acid (EDTA) (is an amino polycarboxylic acid, a colourless and water-soluble solid) as chemical agents that might eliminate inorganic materials on the membrane surface effectively. Indeed, EDTA can produce a strong complex when reacts with Ca^{+2}. Thus, biopolymers associated with Ca^{+2} ions are exchanged by EDTA through a ligand exchange reaction (Al-Amoudi and Lovitt 2007). Kim and Jang (2006) state that the existence of metal ions (e.g. calcium) in membrane bioreactors can be beneficial for the membrane permeation in some membrane processes due to its positive effect on colloid or sludge flocculation.

The mechanism of inorganic fouling involves the crystallisation of the salts ion precipitated from the bulk feed solution and particulate fouling. Lee et al. (1999) point out that there are indeed two ways of crystallisation in the membrane filtration using $CaSO_4$ as the fouling salt: the first way is surface crystallisation (homogenous) and the second one is bulk crystallisation (heterogeneous). Surface crystallisation (homogenous), on the one hand, produces solid crystals directly on membrane surface. It is caused by formation of nuclei on the membrane active sites or impurities on the membrane surface as nucleation sites and lateral growth of crystals. By contrast, bulk crystallisation occurs due to the homogenous or secondary crystallisation in the bulk phase and then deposited to the membrane surface (Okazaki and Kimura 1984; Pervov 1991). These two mechanisms of crystallisation may simultaneously emerge through membrane filtration, and flux decline takes place (Lee and Lee 2000). Figure 3.6 shows the scale formation mechanisms in NF.

Biofouling

Biofouling is generally one of the most common and serious issues for membrane used in many applications such as bioseparation, water and wastewater treatments, membrane bioreactors, reverse osmosis, desalination (Xu et al. 2010; Miura et al. 2007; Kramer et al. 1995; Flemming 1997; Flemming et al. 1997; Baker and Dudly

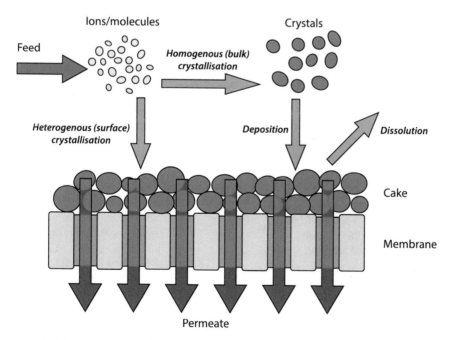

Fig. 3.6 Schematic representation of inorganic fouling mechanisms. Adapted from Shirazi et al. (2010)

1998; Guo et al. 2012; Liao et al. 2004; Ramesh et al. 2006; Liu et al. 2012; Yao et al. 2010, 2011). It appears after organic, inorganic and colloidal fouling.

Many researchers, for example Baker and Dudly (1998), Huang et al. (2012b), Khan et al. (2011), Vrouwenvelder et al. (2011), Baker (2004), Flemming and Schaule (1988) state that membrane biofouling causes a number of serious issues which include

1. A reduction in membrane permeation due to establishment of a gel-like diffusion barrier (e.g. the biofilm) on the membrane surface.
2. An increase in solute concentration polarisation accompanied by lower solute rejection (in RO and NF membranes).
3. An increase in the module differential pressure (ΔP) and feed pressure being required to maintain the same production rate due to biofilm resistance.
4. Biodegradation and/or biodeterioration of the membrane polymer or other module construction materials (e.g. polyurethane-based glue lines).
5. Establishment of concentrated populations of primary or secondary human pathogens on membrane surfaces.
6. Increased energy requirements due to the higher pressure being required to control the biofilm resistance and the flux decline.
7. Shortens membrane lifespan.

As noted by Pang et al. (2005), Wang et al. (2005), biofouling refers to deposition/retention, growth, and metabolism of bacteria cells (marine bacteria, diatoms and green algae) or flocs on the membranes, which is a significant concern in membrane filtration processes. Biofouling is also a major hitch for pressure—driven membranes such as ultrafiltration and microfiltration (used for treating wastewater) because most foulants (colloids and sludge flocs) in MBRs are much larger than the membrane pore size. Kochkodan (2012) defined membrane biofouling as a dynamic process of microbial colonisation and growth, which results in the formation of microbial films onto membrane surface, differentiation of microbial films into mature biofilm and eventually biofilm detachment and dispersal as shown schematically in Fig. 3.7.

Biofouling often makes the membrane surface become non-regenerable and therefore more replacement or cleaning of the membrane is required, which contributes significantly to the application cost (Baker and Dudly 1998). As a matter of fact, biofouling term used to describe many fouling phenomena where biologically active organisms such as fungi, viruses and microorganisms and excreted extracellular biopolymers are involved (Flemming 1997; Flemming et al. 1997; Liao et al. 2004; Ramesh et al. 2006). Biofouling is considered as a common encountered problem of the synthetic polymeric membrane surface because it diminishes the treatment process efficiency and cost effectiveness (Flemming and Schaule 1988).

Membrane biofouling is inherently very complicated than other membrane fouling phenomena because microorganisms can grow, multiply and relocate over time on the surface of membrane. Biofouling has been regarded as a contributing factor to more than 45 % of all membrane fouling phenomena (Komlenic 2010).

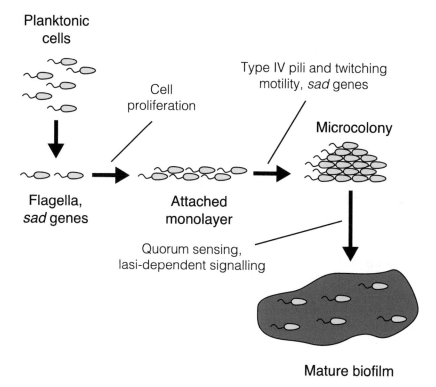

Fig. 3.7 Models of the development of a mature P. aeruginosa biofilm from planktonic cells. Adapted from Kim et al. (2007), Wu (2012)

Membrane biofouling is usually initiated by microbes or bacteria that attach and grow on the surface of the membranes in use (Miura et al. 2007; Vrouwenvelder and Kooij 2001; Baker and Dudly 1998; Flemming 1997; Flemming et al. 1997). Since most conventional membranes are prone to bacterial attachment and growth (Baker and Dudly 1998; Flemming et al. 1997). Baker and Dudly (1998), Chen et al. (2004) point out that membrane biofouling is mainly caused by Corynebacterium, Pseudomonas, Bacillus, Arthrobacter, Flavobacterium and Aeromonas bacterial species and to a minimum extent by fungi such as Trichoderma, Penicillium and other eukaryotic microorganisms. Some microorganisms inherently seem 'sticky' and rapidly tend to adhere to any surface practically, while others respond more slowly and only adhere to certain surfaces after some time. Flemming and Schaule (1988) found that Pseudomonas vesicularis, Acinetobacter calcoaceticus and Staphylococcus warneri have been identified as fast adhering species out of a tap water microflora: then, the first irreversible attachment of cells take places after few minutes of contact between membrane and raw water. If non-starving cells are used, the adhesion process is strongly dependent on the number of cells in suspension with a linearity between the logarithms of numbers of suspended and adhering cells above 10^6 cells/ml up to complete covering of the surface (3×10^7/cm^2). Starving cells do not cover the surface completely but colonise it in an island pattern with free membrane areas in between Flemming and Schaule (1988). As a matter of fact, the differences in microbial adhesion to membrane surfaces over long time may result from biological factors such as the nutritional condition and growth plate of the microbe, but the initial attachment is largely governed by physico-chemical factors such as hydrodynamic conditions (Kang et al. 2004; Schneider et al. 2005) and membrane surface properties (Ridgway et al. 1985).

Many investigators propose that soluble microbial product (SMP) and extra-cellular polymeric substance (EPS) released by bacteria also play important roles in the formation of biological foulants and cake layer on membrane surfaces by altering their physico-chemical characteristics such as hydrophilicity and surface charge (Flemming et al. 1997; Liao et al. 2004; Ramesh et al. 2006; Neu and Marshall 1990). EPS primarily consists of polysaccharides, proteins, glycoproteins, lipoproteins and other macromolecules of microbial origin (Beer and Stoodley 2013). It contains both hydrophobic and hydrophilic sites on its chemical structure, which enable the polymer to settle on both hydrophilic and hydrophobic properties. As noted by Neu and Marshall (1990), the interaction force between the EPS and the membrane surface may be physical (adsorption), chemical (covalent bonding) or electrostatic. EPS account for 50–90 % of the total organic carbon (TOC) of biofilms and can be considered as the primary matrix material of the biofilm.

Membrane biofouling is usually regarded as irreversible process and is very hard to handle due to the self-replicating nature of microbes or foulants (Flemming et al. 1997; Kappachery et al. 2010). Therefore, many techniques have been performed over the last two decades to control membrane biofouling. The common strategy to deter membrane biofouling is often to add biocides or antibacterial agents, such as heavy metals, including zinc, copper, chlorine, silver into the feed stream of the membrane process (Huang et al. 2012b; Zhang et al. 2012; Zhu et al. 2010). The use of chlorine would kill 99.99 % of bacteria in the feed stream, this approach may not be effective to eliminate membrane biofouling because they are still enough bacteria remaining which can migrate to the membrane surface, relocate and multiply rapidly, especially at high feed temperature and in wastewater treatment or desalination where higher contents of nutrients are available in the feed for bacterial growth (Zhu et al. 2010). Besides chlorine, silver is another type of biocide or antibacterial agent, which is extensively used in different applications, despite the fact that the study of silver for the purpose of membrane anti-biofouling has been very limited (Chae et al. 2009; Zhu et al. 2010). Silver nanoparticles have attracted considerable attention due to their catalytic, optical and conducting properties (Koga et al. 2009). It is concluded that the use of silver as an effective antibacterial agent has well known for a long time (Koga et al. 2009; Matsumura et al. 2003). Lee et al. (2007) argue that the grafting of silver nanoparticles onto the surface of polyamide nanofiltration membrane was shown to prevent biofouling problem effectively and also enhance the performance of nanofiltration membrane. Huang et al. (2014) demonstrate that the incorporation of Ag-SiO$_2$ to PES membrane would improve the hydrophilicity of the membrane, which will also enhance the anti-biofouling properties. Apart from silver, zinc has also been used as anti-biofouling membrane (Zhang et al. 2014). Incorporating zinc into polymer membranes would improve antibacterial or anti-biofouling performance. Zhang et al. (2014) concluded that introducing poly (zinc acrylate) (PZA) into PES membrane would be a simple route to improve the anti-biofouling properties of membranes. The biofouled membranes can be visualised using different analytical techniques such as Atomic force microscopy (AFM), Scanning electron microscopy (SEM), Confocal Laser scanning microscopy (CLSM) and direct observation through the membrane (DOTM).

Reversible and irreversible fouling

Irreversible fouling is a kind of fouling that exhibits a marked dependence on the surface membrane properties, whilst reversible fouling is only weakly dependent on the membrane surface chemistry.

Irreversible fouling is caused by strong adherence to the membrane such as pore blocking, gel layer formation or biofilm formation. Huyskens et al. (2008), Meng et al. (2009) point out that irreversible fouling cannot be removed by any methods including chemical method (using sodium hypochlorite (NaOCl) as chemical agent for cleaning). But, many researchers (Huang et al. 2012a, b; Judd 2007; Judd and Jefferson 2003) state that the irreversible fouling can solely be eliminated by chemical reagent but repeated chemical cleaning may reduce the membrane performance, leading to membrane degradation. Reversible fouling, on the contrary, occurs due to external deposition of material (cake filtration) and is mostly removed by hydraulic cleaning such as backwashing or cross-flusing and relaxation. Concentration polarisation and cake layer formation are often considered as important reversible fouling mechanisms. A superior control of both reversible and irreversible fouling would decrease operational costs associated with membrane cleaning, thereby making MBRs more competitive in comparison to conventional wastewater treatment plants (Judd 2006). Huyskens et al. (2008) used a new method to evaluate irreversible and reversible fouling. This method is called the MBR-VFM (VITO Fouling Measurement). This method is a further modification of the VFM, which was developed about 5 years ago as an alternative method to determine the fouling tendency of aqueous feeds in pressure-driven filtration processes. Table 3.3 presents typical ranges of various fouling rates at full-scale plant. As can be seen from the table, each fouling type has a different fouling rate and time frame over which it appears.

Table 3.3 Typical ranges for different fouling rates occurring at full-scale operation (Guglielmi et al. 2007a, b; Kraume et al. 2009; Pollice et al. 2005)

Category	Fouling rate in mbar/min	Time frame
1. Reversible fouling (Cake filtration)	0.1–1	10 min
2. Residual fouling	0.01–0.1	1–2 weeks
3. Irreversible fouling	0.001–0.01	6–12 months
4. Irrecoverable fouling	0.0001–0.001	Years

Colloidal and particle fouling

Colloidal fouling refers to membrane fouling with colloidal and suspended particles in the size range of a few nanometres to a few micrometres. Particulates below this range are considered to be dissolved in water. A wide number of colloidal and particulates foulants with different size range have been studied by many investigators (Potts et al. 1981; Yiantsios et al. 2005). Examples of these foulants include silts and clays, silica salts, humic acid, hydroxides of heavy metals, precipitated crystals and iron, and aluminium compounds (Buffle and Leppard 1995; Buffle et al. 1998). Some of these foulants include

- Microorganisms.
- Biological debris (plant and animal).
- Polysaccharides (gums, slime, plankton, fibrils).
- Lipoproteins.
- Clay (hydrous aluminium and iron silicates).
- Silt.
- Oils.
- Kerogen (aged polysaccharides, marine snow).
- Humic acids, lignins, tannins.
- Iron and manganese oxides.
- Calcium carbonate.
- Sulphur and sulphides.

The increased concentration of the rejected ions at the front of the membrane surface facilities the accumulation of dissolved organic substances, for instance, natural organic matter (NOM) onto colloidal-sized particles (Hong and Elimelech 1997). Colloidal fouling mainly depends on the colloidal particle relative to the membrane pore size.

As stated by Boerlage et al. (2003), silt density index (SDI) and modified fouling index (MFI) are the most widely applied methods in the NF and RO that are used for evaluation of membrane fouling potential generated by dispersed particulate matters (colloids, suspended solids) in the feed. Severe fouling is expected to be high when SDI and MFI are high. SDI and MFI methods use only one number value to evaluate the feed water. However, these fouling indices are based on feed passing through a 0.45-μm microfiltration membrane to measure the fouling potential, they cannot explain the decrease of flux rate effectively because of their limited ability to retain colloids smaller than 0.45 μm which is responsible for membrane fouling. As stated by Brauns et al. (2002), SDI and MFI indices use only one value to predict the fouling propensity of the feed, failing to reflect different fouling mechanisms of RO/NF membranes. Therefore, development of alternative robust and predictable fouling indices has been the subject of numerous research studies. This can be done by modifying fouling index experiments to utilise test membranes with pore size smaller than 0.45 μm. The colloids film can be formed in two stages; internal and external fouling (Czekaj et al. 2000; Visvanathan and Ben aïm 1989) as was discussed earlier in this chapter.

3.1.3 Techniques for the Control of Membrane Fouling in Membrane Bioreactors

As mentioned previously, membrane fouling is a very complex phenomenon in all membrane processes, especially membrane bioreactors. The complexity of membrane fouling means that a variety of approaches will be necessary to decrease its impact (Hilal et al. 2005). These approaches are grouped under six main topics (1) pretreatment of feed, (2) optimization of operating conditions, (3) cleaning procedure, (4) membrane materials/surface modification of membrane, (5) gas sparging and (6) pulsatile. These distinct approaches are discussed in detail in the following sections.

3.1.3.1 Pretreatment of Feed

The effects of membrane fouling can be reduced by feed pretreatment. Feed pretreatment is a crucial step in MBR process. This approach is widely used to eliminate the particulates or macromolecules that cause pore clogging or prevent them from depositing onto the surface of the membrane. Feed pretreatment is an effective approach as it contributes to reducing the consequences of membrane fouling (biofouling) and it involves both physical and chemical approaches. The physical approaches normally include prefiltration or centrifugation process to eliminate the suspended particles that have the tendency to plug the membrane module or attach to the membrane surface. Heat treatment followed by settling is the most common type of physical process. It is used specifically in dairy plant to remove fats and immunoglobulin's from cheese whey prefiltration prior to ultrafiltration (Aptel and Clifton 1986).

In contrast, chemical approaches are another approach, which include different processes such as coagulation, precipitation or flocculation and the use of proprietary chemicals as disinfectants. pH adjustment of the feed causes foulants to further away from their isoelectric points which eventually reduce their propensity to form a gel layer (Peuchot and Aim 1992). Removing any colloids (particulate matter) and therefore preventing particles from depositing on the surface of membranes requires the usage of both chemical and physical processes. As stated by Boissonade et al. (1991), using coagulation agents and flocculation could reduce membrane fouling by the accumulation of the colloidal fraction, therefore, reducing the internal blocking of the membrane. The function of the coagulants is to eliminate the internal blocking of the membrane by settling the colloidal matters and supporting them to form large aggregates in order to produce higher rates of permeation flux. The addition of coagulant can not only improve permeation flux but also enhance the quality of membrane effluent. Four distinct mechanisms of chemical coagulants can impair colloidal particles, which include (i) charge neutralisation, (ii) double layer compression, (iii) enmeshment in a precipitate and (iv) interparticle bridging (Hilal et al. 2005).

A research study was conducted on using coagulants to improve the formation of large aggregates from the initial molecules that would be easily swept of the membrane surface. Two types of coagulant have been used for water and wastewater treatment application such as ferric chloride and alum (also called aluminium sulphate). These coagulants are thoroughly added to significantly reduce membrane bioreactor fouling. As noted by Holbrook et al. (2004), when alum dissolved in water, it forms hydroxide precipitates which adsorb materials such as suspended particles, colloids and soluble organics. In MBR-based trials, the addition of alum resulted in significant decrease of the SMPc [fraction of carbohydrate contained in the sludge solution (mg/gSS)] concentration, along with an amelioration in the hydraulic performance of the membrane. Large microbial flocs are supposed to have a lower impact on membrane fouling because of back transport and shear induced fouling control mechanisms. Therefore, the improvement of permeability or permeate flux in MBRs system is due to the formation of largest flocs. As stated by Lee et al. (2001b) small biological colloids (from 0.1 to 2 μm) have been observed to coagulate or condensed and formed larger particles when alum is added to MBR-activated sludge.

Although highly costly, dosing with ferric chloride ($FeCl_3$) was found to have higher efficiency than that of alum. It was used to remove large-sized organic compounds. In MBRs, zeolites have also implemented and promoted the creation of rigid flocs that have specifically lower fouling resistance. Ferric iron has also been added to MBRs to enhance the production of iron oxidising bacteria, which is responsible for the degradation of H_2S gas. Specific ferric precipitate, for example ferric phosphate and K-jarosite ($K-Fe_3(SO_4)_2(OH)_6$) have been seen to foul the membrane (Park et al. 2005). A recent research study conducted by Itonaga et al. (2004) showed that both suspension velocity and irreversible fouling can be controlled with the addition of iron as a coagulant. Ferric hydroxide flocs have been used as membrane precoating agent in membrane bioreactor process. The results showed that the permeability was higher than that of uncoated MBR system. Also the effluent quality was improved significantly in this study (Zhang et al. 2004). In that study ferric chloride was used in MBR systems, which successfully removed the non-biodegradable organics that aggregated in the MBR systems. This operation also resulted in significant increase in membrane permeation or permeability flux

3.1.3.2 Optimization of Operating Conditions

Improvement of operating conditions at the membrane surface during membrane operation (for example, CFV (cross-flow velocity) and shear stress) is one of the approaches used to mitigate membrane fouling as it leads to an increase in mass transfer coefficient and turbulence, thus weakening concentration polarisation.

Aeration

Many efforts have been conducted on optimisation of air flow rate to reduce the cost factor of energy involved in providing aeration to membranes (Le-Clech et al. 2006b).

Two distinct parameters to limit membrane fouling are the location of aerators and the specific design of airflow patterns. Further enhancement in aeration design implemented by membrane bioreactor manufacturers are usually reported in many different ways (e.g. patent, format) and include cyclic aeration systems (Rabie et al. 2003), and improved aerator systems (Miyashita et al. 2000). Recently, Mayer et al. (2006) outlined an exhaustive comparison of various aeration devices used in tubular membranes. They concluded that complex aeration systems with multiple orifices injecting air homogeneously in the feed flow worked best. The effect of aeration differs from hollow fibre membrane modules to flat plate membrane modules. Bichamber (riser and downcomer) presence in a Kubota MBR has a significant impact in generating high cross-flow velocity (CFV m/s) (Sofia et al. 2004). Higher CFV and lower uplift resistance were induced by uniformly distributed fine air bubbles (produced from a porous media with 0.5 mm holes) in comparison to performances achieved with larger bubbles (from 2 mm hole diffuser) at the same aeration rates. The ratio of air/permeate (m^3/m^3) is a useful parameter to characterise aeration intensity which is needed to obtain a specified amount of treated water. Values given by MBR manufacturers can differ between 24 and 50, depending on membrane modules (flat sheet versus hollow fibre) and the design of MBR tank (membrane and aerobic zone combined into one tank or not) (Tao et al. 2005). Preliminary work implemented in Singapore on large-scale MBRs showed the original ratios to be quite conservative, since it was possible to decrease them (down to 56 % of their original value) without significantly increasing fouling (Tao et al. 2005). Numerous research studies have been conducted to increase the critical flux in submerged MBR by alternating the aeration rates (Howell et al. 2004). During high throughput period, aeration was increased and dropped to lower values for the low throughput period, to reduce membrane fouling. This technique was used also to reduce energy consumption. As stated by Choi et al. (2005), when pseudosteady state is achieved, the tangential shear was found to have no effect on permeation flux decline in cross-flow MBR device. Increasing CFV has no capability to reduce fouling when the layer begins to deposit and dominate the permeation flux behaviour. Flux decline was caused predominantly by reversible fouling when CFV was not present, when CFV was present slightly higher irreversible fouling was observed (Choi et al. 2005). Intermittent operation of aeration has also been reported for (de) nitrification MBR systems (Yeom et al. 1999; Nagaoka and Nemoto 2005). In this uncommon scheme, a single tank MBR was used for anoxic and aerobic biological degradation; filtration process is executed during the aerobic phase to take advantage of the antifouling properties of the air scouring. As soon as air sparging ceased, severe fouling was observed when researchers tested intermittent aeration, therefore they concluded that this kind of test is not useful (Jiang et al. 2005; Psoch and Schiewer 2005). Other investigators have indicated that using intermittent bubbling is an efficient way to control fouling (Judd et al. 2006; Fane 2005). Pulsing air at a frequency of 1 s on/1 s off prompted further enhancement in operating flux ranging from 20 to 100 % and was noted to be a more effective method than using lower frequencies (5–10 s on/5–10 s off) which is commonly used industrially (Judd et al. 2006).

However, such system may need the operation of robust activators and valves at these high frequencies and may not be economically practical.

Other operating conditions

Le-Clech et al. (2006b) defined solid retention time (SRT) as one of the main operating parameters defining the properties of the biomass suspension and its fouling tendency. It also influences membrane bioreactor performance, especially in the control of fouling. With numerous research studies explaining the relationship between SRT and concentrations of both extracellular polymeric substances (EPS) and soluble microbial product (SMP), it is evident that overall performance of the MBRs is almost related to SRT value. To enhance filtration performance a long SRT is used, this also minimises SMP and EPS formation by producing starved conditions (Judd 2006). However, using a very long SRT, severe membrane fouling is to be expected as a consequence of the accumulation of MLSS or sludge (filamentous) production.

Moreover, the performance of MBR is reduced due to low biomass when the SRT is extremely too short (down to 2 days) (Meng and Yang 2007). Meng and Yang (2007) stated that higher ratio of Food/Microorganism can also increase the concentration of EPS because of high food utilisation by biomass.

Further optimizations of operating conditions through reactor design have been investigated and included the addition of a spiral flocculator (Guo et al. 2004), vibrating membranes (Genkin et al. 2005), helical baffles (Ghaffour et al. 2004), suction mode (Kim et al. 2004) and high performance compact reactor (Yeon et al. 2005), novel types of air lift (Chang and Judd 2002), porous and flexible suspended membrane carriers (Yang et al. 2006a) and the sequencing batch MBR (Zhang et al. 2006a), for example. Lastly, the design of membrane modules remains as another vital factor in the optimization of the MBR operation, and more accurately, the use of air sparging method. Higher permeability was obtained, when a specific module was designed with air bubbles confined in close proximity to the hollow fibre (Ghosh 2006).

Sustainable flux

The energy requirement for operation is a potential weakness for the future development of the MBR systems. In comparison to the conventional activated sludge systems, it is acknowledged that the energy usage of MBRs is still higher because of the need to alleviate membrane fouling by different techniques (Le-Clech et al. 2006b). At the end of the day, MBRs can be economically feasible only if it produces a reasonable flux rate without significant fouling. When permeation rate and membrane fouling reduce simultaneously, most MBR systems operate at low fluxes to alleviate excessive membrane fouling. Generally, sustainable flux can be defined as subcritical flux by default. Though, in MBR it can be defined as the flux where the rate of transmembrane pressure (TMP) rises gradually at an allowable rate, removing the need for chemical cleaning (Ng et al. 2005). Before chemical cleaning is required, both the rate of TMP increase and the period of filtration are left to the operator's discretion, and hence a more detailed definition of sustainable flux cannot be possible. While vital flux was commonly determined during short-term experiments, sustainable flux can only be assessed through longer filtration periods. In sustainable flux systems, the value of flux is vital, but the methods used to maintain and achieve the given flux value are also of great importance.

3.1.3.3 Cleaning Procedures

Cleaning process is paramount step towards regenerating the membranes, making them very active for usage or application. Cleaning procedure has to be done when the flux is slightly reduced and transmembrane pressure is increasing drastically. Cleaning procedure is usually performed in three different forms; physical, chemical and combination of physical and chemical cleaning. Physical and chemical cleaning can remove foulants (microbial flocs) from membrane surface through backwashing, relaxation, hydraulic scouring and chemical reactions.

Physical cleaning

In MBR systems, physical cleaning techniques are used to recover membrane permeability significantly. It mainly includes two processes: the first process is relaxation process (occurs where the filtration process is ceased) and the second one membrane backwashing (where the effluent is delivered back through the membrane into the feed channel to remove the deposited particles on the surface of membrane) (Le-Clech et al. 2006b). Relaxation and backwashing techniques have been used in membrane bioreactor systems as standard operating techniques to alleviate membrane fouling; although potent backwashing is not a suitable option for flat plate submerged membranes.

Backwashing (Backflushing or water rinsing) was used to remove reversible fouling effectively, remove contaminants aggregated on the membrane surface and also allowed efficient flux recovery. Frequency, duration and their ratio are the key parameters in the design of backwashing. Jiang et al. (2005) state that using less frequent and longer backwashing duration (600 s filtration/45 s backwashing) has been found to be more effective than using more frequent and less backwashing duration (200 s filtration/15 s backwashing) (Jiang et al. 2005). In another research study, using backwash duration (between 8 and 16 min) was found to be very effective in removing membrane fouling than using either the aeration intensity $(0.3–0.9 \text{ m}^3/\text{m}^2 \text{ h})$ or backwash duration (25–45 s) (Schoeberl et al. 2005) for hallow fibre immersed membrane bioreactor systems. Whilst more fouling is anticipated to be eliminated in more frequent, longer, and stronger backflushing duration, possible modification or alteration is required, exploring to minimise energy consumption. This has been achieved by designing a generic control system which automatically optimised the backflush duration according to the monitored value of transmembrane pressure (Smith et al. 2005). However, a large number of research studies have not taken into consideration the loss of productivity that results from the use of permeation flux during the backwashing.

Obviously, the antifouling operation affects operating costs, as more energy is needed to achieve a suitable pressure for flow reversal. Furthermore, between 5 and 30 % of the permeation flux produced is thoroughly used in this process. In comparison to hollow fibre MBR, flat sheet membrane bioreactor module has achieved a slightly higher overall permeate flux when operating the membrane constantly at low flux (Judd 2002). In this example, flat sheet membranes (which cannot be backwashed) were operated constantly with flux ranging between 20 and 27 LMH. The hollow fibre membrane bioreactor module on the other hand was operated at higher flux, ranging from 23 to 33 LMH but with 25 % of the permeate product being recycled for backwashing (45 secs of backwashing after every 600 secs of operation).

As noted by (Sun et al. 2004), air can be used to affect backflushing. Approximately 400 % increase in the flux has been achieved from continuous operation mode using an air backflush. 15 min of air backwash was needed for every 15 min of filtration to obtain this result (Visvanathan et al. 1997). Whilst, air backwashing is more effective method used to recover membrane permeability. However, air backflushing may present some potential issues of membrane which become rewetting and breakable (emrittlement).

Membrane relaxation (so-called non-continuous process of the membrane or batch process) is intermittent cessation of permeation for flux recovery if the membrane is submerged, and scoured with air when the permeation is ceased. It improves the productivity of membrane significantly. Under this operation, in nature, back transport of foulants (microbial flocs) is enhanced as non-irreversibly attached foulants, which can diffuse away from the surface of membrane via concentration gradient improved by the shear generated by air scouring (Hong et al. 2002; Chua et al. 2002). Comprehensive studies of the behaviour of transmembrane pressure during this operation has been showed that although the rate of fouling is generally higher than for continuous filtration, membrane relaxation allows filtration to be kept for longer period of time before the need for intense chemical cleaning (Ng et al. 2005). Although some research studies have mentioned that this operation may not be feasible economically for large-scale MBRs (Hong et al. 2002), productivity analyses and further cost are possibly required to compare this method against backwashing process.

Recent research studies evaluating another technique to control membrane fouling have been tended to merge relaxation with backwashing for optimum results (Zhang et al. 2005; Vallero et al. 2005). Combination of relax/permeate and backflush/permeate can reduce chemical cleaning and shorten membrane lifespan (Zsirai et al. 2012).

To sum up, physical cleaning only eliminates the coarse solid or cake layer from the membrane surface, whilst chemical cleaning eliminates the flocs. Physical cleaning can also eliminate the strong matters that attach on the surface of membrane. The energy consumption for physical cleaning up to 30 % of the (permeate) must be taken into careful consideration.

Chemical cleaning

As expected, the effectiveness of membrane backwashing and relaxation tend to decrease with operation time as more irreversible fouling aggregates on the membrane surface. Thus, in comparison to the physical cleaning strategies, various chemical cleaning strategies have been recommended or proposed (Le-Clech et al. 2006b). They include

- Enhanced backwash chemically (this can be done each day).
- Maintenance cleaning with higher concentration of chemical agent (this should be done weekly).
- Intensive (or recovery) chemical cleaning (this must be done once or twice a year). This should be done if the membrane permeation is no longer sufficient. This is designed to remove irreversible fouling.

Maintenance cleaning is used to recover membrane permeability, and to reduce the frequency of intensive cleaning. Chemical cleaning controls membrane fouling especially, scaling, organic fouling and biofouling which is not removed by physical cleaning. In general, exhaustive cleaning is conducted when further filtration is no longer sustainable because of higher transmembrane pressure (TMP). Each of the four main MBR suppliers (Kubota, Memcor, Mitsubishi and Zenon) proposes their own chemical cleaning recipes, which differ mainly in terms of concentrations and methods as presented in Table 3.4.

Normally, sodium hypochlorite (0.1–0.5 wt%) is the prevalent chemical agent used to remove organic fouling and biofouling effectively, whilst, citric acid is used to remove inorganic scalants. Sodium hypochlorite (NaOCl) and hydrogen peroxide (H_2O_2) are the most common oxidant agents, which function through oxidation and disinfection. (NaOCl) and (H_2O_2) are used to hydrolyse the organic molecules, and therefore slacken the biofilm and particles that sticked to the membranes. The influences of cleaning chemical agents like NaOCl on microbial community have

Table 3.4 Intensive chemical cleaning protocols for four MBR suppliers (Le-Clech et al. 2005a)

MBR Suppliers	Type	Chemicals	Concentration (%)	Protocols
Mitsubishi	CIL	NaOCl Citric acid	0.3 0.2	Backflow through membrane (2 h) + soaking (2 h)
Zenon	CIP	NaOCl Citric acid	0.2 0.2–0.3	Backpulse and recirculate
Memcor	CIP	NaOCl Citric acid	0.01 0.02	Recirculate through lumens, mixed liquors and in-tank air manifolds
Kubota	CIL	NaOCl Oxalic acid	0.5 0.1	Backflow and soaking (2 h)

CIL cleaning in line where chemical solutions are generally backflow (under gravity) inside the membrane. *CIP* cleaning in place where membrane tank is isolated and drained; the module is rinsed before being soaked in the cleaning solution and rinsed to remove excess of chlorine

been conducted for modelled MBR systems (Lim et al. 2004). They postulated that the performance of organic degradation of the microbial community in the occurrence of NaOCl was hampered. They also suggested that adding additional amounts of NaOCl caused an inhibition of organic degradation and cell lysis.

It is also usually for MBR manufactures to adjust specific protocols for chemical cleaning (i.e. cleaning frequencies and chemical concentrations) and for individual facilities (Tao et al. 2005; Kox 2004; Le-Clech et al. 2005a). It also has been mentioned that the level of pollutants (measured as TOC (total organic carbon)) in the permeate (effluent) rises just after the chemical cleaning episodes (Tao et al. 2005). This is important for MBRs utilised in reclamation process trains (i.e. upstream of RO). Until now, no systematic research studies have been conducted on cleaning procedures (Liao et al. 2004). This is probably due to the site-specific nature of the membrane bioreactors fouling.

Maintenance cleaning takes up to 30–120 min for a cycle to be completed. Usually, it is implemented every 3–7 days (approximately 1 week), using NaOCl as chemical agent. The concentration of NaOCl should be moderate (200–500 mg/l) or 0.01 wt% NaOCl for aerobic membrane bioreactor systems. Recovery cleaning uses rather higher reagent concentrations of 0.2–0.5 wt% NaOCl coupled with 0.2–0.3 wt% citric acid or 0.5–1 wt% oxalic acid (Judd 2011).

3.1.3.4 Optimization of Membrane Characteristics

It is another approach used to avoid fouling in MBRs system and increased membrane permeability. Chemical modification of the membrane surface has been found to improve antifouling performance efficiently and effectively. When hydrophobic membranes are used in the MBR, more severe fouling is expected. Therefore, numerous research studies have been conducted on hydrophilisation (make the membrane more hydrophilic) of membrane. Yu et al. (2005a, b) modified polypropylene (PP) hollow fibre membranes using NH_3 and CO_2 plasma treatments. In both research studies, X-ray photoelectron spectroscopy and scanning electron microscopy were conducted to characterise the morphology and chemical structure of the modified PP membrane surface. Introducing polar groups (from oxygen and nitrogen) into the membrane surface, membrane hydrophilicity increased significantly and the modified membranes showed excellent separation performances and flux recovery ratio in comparison to the neat membranes (Le-Clech et al. 2006b). A detail overview will be provided in chapter four on surface modification.

3.1.3.5 Gas Sparging/Air Sparging

Gas or air sparging is another technique used to generate high surface shear at membrane surfaces, reducing membrane fouling and significantly improving the performance of some membrane processes (i.e. UF or MF) due the advantageous bubbling method to control the concentration polarisation and cake layer deposition (Cui et al. 2003). In MBRs processes, air sparging method is usually employed because the MBRs required aeration system to provide oxygen as a substrate for biomass suspension and microorganism. Oxygen is very necessary in MBRs to keep the existing biomass alive and to degrade the biodegradable pollutants. By directly injecting air in the concentrate section during the filtration process, air sparging creates an intermittent two-phase flow (gas/liquid) along a membrane surface, leading to reduce fouling to a minimum, enhanced flux and increase surface shear. Ducom et al. (2002) stated that air sparging is very effective method used with both tubular membranes and hollow fibre membranes for Escherichia coli suspensions. Air sparging effectively enhanced the membrane permeation in microfiltration and ultrafiltration for a wide range of applications, particularly for particles or solutes in water. However, air sparging technique is not as efficient as the use of turbulence promoters (e.g. spiral wire, metal grill, static rods) due to the handling problem of gas injected into the membrane (Wakeman and Williams 2002b).

The method of injecting air into the feed stream to mitigate colloidal or particulate membrane fouling in organic hollow fibres has been widely reported in the literature. The efficiency of a two-phase flow (gas/liquid) was assessed to prevent cake layer formation; while the effects of injecting air on permeation flux at different velocities were scrutinised by Cabassud et al. (1997). The results showed that the permeation flux (in case of non-injecting air in the feed stream) decreased with time during filtration operation. However, during the first few minutes, the permeation flux showed a significant decrease due to the depositing particle within the membrane, therefore, the added flow resistance. In contrast, the permeation flux (in case of injection air in the feed stream) decreased with time, but the rate of flux decreasing was very low. In case of using higher air velocity, the permeate flux experienced a higher level. Furthermore, increasing air velocity results in particle deposit formation with less mass transfer resistance. Also, the flow of an intermittent gas seems to be effectively less than a steady one in similar experimental operations. It was well documented that there is the evidence of change of fouling layer or cake structure with the application of gas sparging to the microfiltration of particles. This affects its specific resistance by reducing it (Cabassud et al. 1997). An improved flux over time was a result of alterations in the cake structure or fouling layer with the gas-sparged ultrafiltration of clay suspension.

The effects of injecting air into the feed side of flat sheet membrane modules on protein transmission and permeation flux were scrutinised using four foulant models such a human immunoglobulin G (IgG), bovine serum albumin (BSA), human serum albumin (HSA) and lysozyme (Lys) as test media (Li et al. 1998). Polysulfone and polyethersulfone were the two membranes used in this study. The results showed that gas sparging enhanced membrane flux up to 50 % but protein reduction/retention was reduced significantly. It was found that at TMP of approximately 50 kPa, with using a mixture of BSA and Lys, the gas injection increased membrane flux by only 10 % but had a more powerful effect on protein separation. In another research study, the flux was increased between 60 and 270 %

due to gas sparging obtained on ultrafiltration of (BSA and dextran) solutions and also in the microfiltration of yeast suspensions using tubular membranes (Cui and Wright 1994; Cui 1993; Lee et al. 1993; Imasaka et al. 1993; Cui and Wright 1996).

A method of incorporating gas into the feed stream as a process of separating the dense cake layer on the membranes was scrutinised using two-phase (gas–liquid) cross-flow ultrafiltration in the downward flow condition using a tubular membrane module (Cui and Wright 1996). The module was vertically installed, with both the gas bubbles and feed solution flowing downward in the membrane. The membrane permeation achieved in this method was in comparison to that in conventional single-phase ultrafiltration and gas-sparged upward cross-flow operations. The operating parameters which were scrutinised with dextran solutions (protein model) included gas flows, liquid flows, transmembrane pressure (TMP) and feed concentrations. The results showed that the permeation flux was increased up to 320 %. It was noted that the addition of gas in both upward and downward flows resulted in significant enhancement in permeation flux. This confirmed the technique to be very useful in deterring concentration polarisation and hampering flux in comparison to any other technique known at that time. The improvement in permeation flux was more obvious when the concentration polarisation was very severe, indicating the validity of the disruption. Using a low flow rate of gas sparging results in improved ultrafiltration while the introduction of the bubbles enhanced an early transition from laminar flow to turbulent flow. The bubbles could represent as static baffles or slow moving baffles to decrease the rate of flow gas, depending on the process conditions.

Air sparging is a strategy that is employed to enhance the efficiency of backwashing in a dead-end fibre module. It was also investigated and effectively confirmed for the removal of the cake layer formed during the filtration of bentonite suspensions (Serra et al. 1999; Ducom et al. 2002). The rinse phase efficiency was significantly enhanced after the introduction of air, acting as a piston to flush out the major part of the free volume in the module.

The combination of air sparging and back shock enhances the productivity as less permeation flux is consumed during backwash process. On the other hand, the decline of permeation flux caused by membrane fouling usually cannot fully be refilled through gas sparging because the internal pore fouling is not reversed by surface shear. Therefore, bubbling or gas sparging can efficiently be used for membrane processes, especially microfiltration and ultrafiltration processes. It can also be used within membrane modules or in submerged membrane bioreactors systems as they have been proved to improve the rates of permeation flux and reduce membrane fouling to a minimum.

Air sparging performance was examined on submerged MBR systems used for wastewater treatment with emphasis on two techniques. The first technique included the use of air injection into membrane tube channels to allow circulation of mixed liquor in the bioreactor, and the second one used periodic air jets into the membrane tube. The results achieved from both the techniques showed that the cake layer was sufficiently detached or separated from the membrane due to use air injection.

Another application of using air sparging for improvement of membrane flux or permeation flux in nanofiltration membranes was studied using droplet suspensions in water (Ducom et al. 2002). The process was examined on a flat sheet membrane for two feeds: stabilised and non-stabilised oil–water emulsions. The focus of the study was based on using nanofiltration membranes in the industrial section including treatment of cutting oil fluids, which are encountered in metal manu-facturing, metal working, and processing. The function of this study was to remove the dissolved organic components from the aqueous phase and also to separate the oily dispersed phase from the membrane. The experiments were conducted to achieve two objectives: the first one was to achieve an oily phase as concentrated in oil as possible so to be able to reuse the oil; whilst the second one was to obtain an aqueous phase in accordance with the regulation standards for industrial wastew-ater. The preliminary results of the experiments showed that the injection of air at very high velocities did not change the permeability or permeate flux to pure water.

Air sparging with non-stabilised oil–water emulsions also showed lower filtra-tion capability due to the formation of an oil layer at the surface of membrane as the membrane hydrophobicity increases in the absence of surfactants. It was noted that in both cases, a significant increase in permeation flux was achieved with air sparging. This was a result of the capability of air bubbles for distorting the oil layer over the surface of membrane.

Air sparging has also been proved to be more effective and very efficient with both tubular membranes and hollow fibre membranes for *E. coli* suspensions, dextran or albumin solutions (Bellara et al. 1996), clay suspensions, clay particles (Mercier et al. 1997; Laborie et al. 1998) and natural surface waters (Cabassud et al. 2001).

3.1.3.6   Pulsatile Flow

The use of pulsatile flow has been equipped towards producing unsteady flows and oscillations. Oscillations and unsteady flows are obtained by introducing pulsations into the feed (filtrate) and permeate channels (Wakeman and Williams 2002b). The filtration performance of yeast cell harvesting was significantly improved using oscillatory flow mixing in both flat sheet and tubular membranes. The advantage of implementing the method was a sevenfold increase in the membrane permeation. However, studies are still in progress to determine the effects of frequency and amplitude (Howell et al. 1993; Wu et al. 1993). Many researchers noticed that the enhancement in membrane permeation was roughly 300 % increase when they used periodically spaced, doughnut-shaped baffles in ultrafiltration tubes and pulse flows. Pulsing the flow results increased the permeability (flux) by 50 % (Finnigan and Howell 1989; Gupta et al. 1993), and 20 % when tangential inlet ports are used to induce helical flow (Holdich and Zhang 1992).

A variation of the pulsating flow technique was also studied using as an inter-mittent jet of fluid at the inlet to a tubular filter. The results showed that the permeation fluxes (2.5 times higher) were higher than those achieved without intermittent jets (Arroyo and Fonade 1993). It was documented that a further variation of the pulsating flow technique was also possible using a flexible tube at the inlet to a ceramic with the tube subjected to alternating pressure. The results illustrated that the fluxes were increased to 60 % in this study (Bertram et al. 1993).

Researchers have stated that using a periodically spaced baffle with pulsatile flow is effective method in significantly enhanced microfiltration. Whilst, other researchers have verified the effectiveness of pulsatile flow techniques for enhancing permeation flux via using a rotating perforated disc placed in front of the entry section of a bundle of tubular membranes (Wenten 1995; Mackay et al. 1991; Wang et al. 1994; Spiazzi et al. 1993; Gupta et al. 1992; Belfort 1989). This would result in temporarily increase in velocity in the different tubes. It was also showed that the rate of pulsatile flow increased the rate of wall shear as velocity increased at high frequencies (Winzeler and Belfort 1993).

Another technique that was evaluated was the air flush technique, which involves using an intermittent two-phase (gas or liquid) flush via the tube to shift and remove the deposition of cake layer. It was concluded that air flush technique [using two-phase (gas or liquid)] is more efficient technique than using a single-phase liquid flush (Verbeck et al. 2000).

3.1.3.7 Bubbling

As mentioned before, aeration in MBRs consumes more energy, leading to higher operating costs for all the MBRs in common use. It also reduces membrane fouling to a minimum. Therefore, numerous methods have been outlined to enhance the hydrodynamic properties in different membrane modules. Bubble inducing surface shear is a major technique to control the higher costs of aeration and membrane replacement which is related directly to membrane fouling (Cui et al. 2003; Drews et al. 2010; Samir et al. 1992; Martinelli et al. 2010; Ndinisa et al. 2006; Phattaranawik et al. 2007; Taha and Cui 2002; Yamanoi and Kageyama 2010).

A number of various two-phase flow models from bubble flow to slug flow can be induced by various bubbling regimes. It is evident that slug bubbles have desirable hydrodynamics properties, improve membrane permeation (effluent) and improve selectivity in different membrane processes and for different membrane modules (Cabassud et al. 2001; Cheng and Li 2007; Ducom et al. 2002; Essemiani et al. 2001; Li et al. 1997, 1998; Mercier et al. 1997; Taha and Cui 2002; Willems et al. 2009). Although the membrane plants have different economies of scales, the energy required of membrane bioreactor systems in treatment plants of municipal wastewater is higher (about 2–4 times) in comparison with conventional activated sludge process (CASP) (Gil et al. 2010; Verrecht et al. 2008). Zhang et al. (2009) stated that frequency and bubble size have a powerful impact on the hydrodynamic properties in slug bubbling flat sheet membrane bioreactor modules. Traditionally, the need is for some high areation intensity in order to induce scouring of the membrane while also providing oxygen sufficiently to the biomass. In comparison to free bubbling, periodic slug bubbles in flat sheet membrane bioreactor systems are economically and effectively improve the mass transfer coefficient, inducing higher wall shear stress while consuming only a very modest amount of air. One can envisage meeting the biological requirements via fine bubble aeration coupled with the use of periodic slug bubbling to control membrane fouling in membrane bioreactor systems. This combined aeration technique will reduce the requirement of aeration and energy in MBRs. Zhang et al. (2011c) concluded that slug bubbles showed excellent antifouling properties in flat sheet membrane bioreactor systems under both short-term and long-term flux operations. In short-term operation, high flux was expected 40 L m^{-2} h^{-1} for 36 h operation. In short-term filtration operation, high flux was expected (40 L m^{-2} h^{-1}) for 36 hr operation. In contrast, in long-term filtration operation, moderate flux was achieved (initial flux was 24 L m^{-2} h^{-1}) for 14 days operation. At low density (2.5 L/min), the slug bubbling reduces the occurrence of irreversible membrane fouling effectively in the initial step and also reduces reversible fouling drastically during long-term filtration operation. Conversely, in free bubbles, the fouling, which is accumulated at the open stage was greater (not completely cleaned) when aerated with free bubble at the same low density. This will lead to an increase in the formation of cake layer when the MBR is preceded.

With increasing consideration being given to energy consumption, slug bubbling in flat sheet membrane bioreactor systems seemed to be energy saving should be an attractive option to free bubble.

3.2   Conclusions

In this chapter, membrane fouling is a repugnant problem in all membrane processes especially, in membrane bioreactors, where the efficacy of the process is restrained by the aggregation of materials on the membrane surface or within the membrane pores. Membrane fouling was thoroughly addressed. Depending on the specific membrane process, membrane fouling can be classified into organic fouling, inorganic fouling, biofouling, particulate or colloidal fouling, reversible and irreversible fouling and removable or irremovable fouling. Organic, inorganic and biofouling occur simultaneously during filtration process and are merged with each other. The interaction between different types of foulants leads to decline permeation flux of the membrane. A number of methods have been used to mitigate membrane fouling to a great extent. Using absorbents or coagulants can reduce the internal clogging of the membrane or reduce the amounts of solute in the solution, and improve the flocculation ability of flocs, which cannot improve membrane permeation but also notably enhance the quality of membrane. Air sparging is an effective method used to remove cake layer from the membrane surface and control concentration polarisation. Hydrodynamics conditions is one of the efficient methods in controlling membrane fouling, any additional improvement would be helpful in mitigating membrane fouling. For example, merging aeration system with membrane module design with CFD simulators might be very effective in enhancing of hydrodynamic conditions. Also operation below the critical flux is very efficient way to avert fouling within a specific filtration system. Physical and chemical cleaning can recover membrane permeation greatly, therefore enhancing antifouling property of membrane.

References

Ahmad AL (1997) Electrophoretic membrane cleaning in dead-end ultrafiltration processes. In: Regional symposium on chemical engineering, UTM, Johor Bharu, pp 281–288

Ahmed A (1997) Electrophoteric membrane cleaning in dead end ultrafiltration process. Paper presented at the regional symposium of chemical engineers, UTM

Al-Amoudi AS, Farooque AM (2005) Performance restoration and autopsy of NF membranes used in seawater pretreatment. Desalination 178(1–3 SPEC. ISS.):261–271. doi:10.1016/j.desal.2004.11.048

Al-Amoudi A, Lovitt RW (2007) Fouling strategies and the cleaning system of NF membranes and factors affecting cleaning efficiency. J Membr Sci 303(1–2):4–28

Ang WS, et al (2011) Fouling and cleaning of RO membranes fouled by mixtures of organic foulants simulating wastewater effluent. J Membrane Sci 376(1–2):196–206

Aptel P, Clifton M (1986) Ultrafiltration. In: Bungay MP, Lonsdale HK, de Pinho MN (eds) Synthetic membranes: science engineering and application. D Reidel Publishing Co, pp 283–288

Arroyo G, Fonade C (1993) Use of intermittent jets to enhance flux in crossflow filtration. J Membr Sci 80(1):117–129. doi:10.1016/0376-7388(93)85137-L

Asatekin A, Kang S, Elimelech M, Mayes AM (2007) Anti-fouling ultrafiltration membranes containing polyacrylonitrile-graft-poly(ethylene oxide) comb copolymer additives. J Membr Sci 298(1–2):136–146

Bai R, Leow HF (2002a) Microfiltration of activated sludge wastewater-The effect of system operation parameters. Sep Purif Technol 29(2):189–198

Bai R, Leow HF (2002b) Modeling and experimental study of microfiltration using a composite module. J Membr Sci 204(1–2):359–377

Baker RW (2004) Membrane technology and applications, 2nd edn. Wiley, England

Baker R (2012) Membrane technology and application, 3rd edn. Wiley, UK

Baker J, Dudly L (1998) Biofouling in membrane systems—a review. Desalination 118:81–89

Belfort G (1989) Fluid mechanics in membrane filtration: recent developments. J Membr Sci 40 (2):123–147. doi:10.1016/0376-7388(89)89001-5

Bellara SR, Cui ZF, Pepper DS (1996) Gas sparging to enhance permeate flux in ultrafiltration using hollow fibre membranes. J Membr Sci 121(2):175–184. doi:10.1016/S0376-7388(96)00173-1

Bertram CD, Hoogland MR, Li H, Odell RA, Fane AG (1993) Flux enhancement in crossflow microfiltration using a collapsible-tube pulsation generator. J Membr Sci 84(3):279–292. doi:10.1016/0376-7388(93)80023-Q

Boerlage SFE, Kennedy M, Aniye MP, Schippers JC (2003) Applications of the MFI-UF to measure and predict particulate fouling in RO systems. J Membr Sci 220(1–2):97–116

Boissonade G, ORichaud, Milisic V (1991) Couplage d'un traitement biologique et d'un procede de separation par membrane. Proc Recents Progres en Genie des Procedes Compiegne 5:117–122

Bouhabila EH, Aïm RB, Buisson H (1998) Microfiltration of activated sludge using submerged membrane with air bubbling (application to wastewater treatment). Desalination 118(1–3):315–322

Brauns E, Van Hoof E, Molenberghs B, Dotremont C, Doyen W, Leysen R (2002) A new method of measuring and presenting the membrane fouling potential. Desalination 150(1):31–43

Brindle K, Stephenson T (1996) The application of membrane biological reactors for the treatment of wastewaters. Biotechnol Bioeng 49(6):601–610

Buffle J, Leppard GG (1995) Characterization of aquatic colloids and macromolecules. 1. Structure and behavior of colloidal material. Environ Sci Technol 29(9):2169–2175

Buffle J, Wilkinson KJ, Stoll S, Filella M, Zhang J (1998) A generalized description of aquatic colloidal interactions: the three-culloidal component approach. Environ Sci Technol 32 (19):2887–2899. doi:10.1021/es980217h

Cabassud C, Laborie S, Lainé JM (1997) How slug flow can improve ultrafiltration flux in organic hollow fibres. J Membr Sci 128(1):93–101. doi:10.1016/S0376-7388(96)00316-X

Cabassud C, Laborie S, Durand-Bourlier L, Lainé JM (2001) Air sparging in ultrafiltration hollow fibers: relationship between flux enhancement, cake characteristics and hydrodynamic parameters. J Membr Sci 181(1):57–69. doi:10.1016/S0376-7388(00)00538-X

Chae SR, Ahn YT, Kang ST, Shin HS (2006) Mitigated membrane fouling in a vertical submerged membrane bioreactor (VSMBR). J Membr Sci 280(1–2):572–581. doi:10.1016/j.memsci.2006. 02.015

Chae SR, Wang S, Hendren ZD, Wiesner MR, Watanabe Y, Gunsch CK (2009) Effects of fullerene nanoparticles on Escherichia coli K12 respiratory activity in aqueous suspension and potential use for membrane biofouling control. J Membr Sci 329(1–2):68–74

Chang IS, Judd SJ (2002) Air sparging of a submerged MBR for municipal wastewater treatment. Process Biochem 37(8):915–920. doi:10.1016/S0032-9592(01)00291-6

Chang IS, Kim SN (2005) Wastewater treatment using membrane filtration—effect of biosolids concentration on cake resistance. Process Biochem 40(3–4):1307–1314. doi:10.1016/j.procbio. 2004.06.019

Chang IS, Lee CH (1998) Membrane filtration characteristics in membrane-coupled activated sludge system—the effect of physiological states of activated sludge on membrane fouling. Desalination 120(3):221–233

Chang IS, Clech PL, Jefferson B, Judd S (2002) Membrane fouling in membrane bioreactors for wastewater treatment. J Environ Eng 128(11):1018–1029

Cheng TW, Li LN (2007) Gas-sparging cross-flow ultrafiltration in flat-plate membrane module: Effects of channel height and membrane inclination. Sep Purif Technol 55(1):50–55. doi:10. 1016/j.seppur.2006.10.026

Chen CL, Liu WT, Chong ML, Wong MT, Ong SL, Seah H, Ng WJ (2004) Community structure of microbial biofilms associated with membrane-based water purification processes as revealed using a polyphasic approach. Appl Microbiol Biotechnol 63(4):466–473. doi:10.1007/s00253-003-1286-7

Cho BD, Fane AG (2002) Fouling transients in nominally sub-critical flux operation of a membrane bioreactor. J Membr Sci 209(2):391–403

Cho J, Song KG, Hyup Lee S, Ahn KH (2005a) Sequencing anoxic/anaerobic membrane bioreactor (SAM) pilot plant for advanced wastewater treatment. Desalination 178(1–3 SPEC. ISS.):219–225. doi:10.1016/j.desal.2004.12.018

Cho JW, Song KG, Ahn KH (2005b) The activated sludge and microbial substances influences on membrane fouling in submerged membrane bioreactor: unstirred batch cell test. Desalination 183(1–3):425–429

Choi H, Zhang K, Dionysiou DD, Oerther DB, Sorial GA (2005) Effect of permeate flux and tangential flow on membrane fouling for wastewater treatment. Sep Purif Technol 45(1):68–78. doi:10.1016/j.seppur.2005.02.010

Chon K, Cho J, Shon HK (2013) Fouling characteristics of a membrane bioreactor and nanofiltration hybrid system for municipal wastewater reclamation. Bioresource Technol 130:239–247

Chua HC, Arnot TC, Howell JA (2002) Controlling fouling in membrane bioreactors operated with a variable throughput. Desalination 149(1–3):225–229. doi:10.1016/S0011-9164(02) 00764-6

Cicek N, Franco J, Suidan M, Urbain V, Manem J (1999) Characterization and comparison of a membrane boireactor and a conventional activated-sludge system in the treatment of waste water containing high molecular weight compounds. Water Environ Res 71(1):64–70

Combe C, Molis E, Lucas P, Riley R, Clark MM (1999) The effect of CA membrane properties on adsorptive fouling by humic acid. J Membr Sci 154(1):73–87

Costa AR, de Pinho MN, Elimelech M (2006) Mechanisms of colloidal natural organic matter fouling in ultrafiltration. J Membr Sci 281(1–2):716–725

Cui ZF (1993) Experimental investigation on enhancement of crossflow ultrafiltration with air sparging

Cui ZF, Wright KIT (1994) Gas-liquid two-phase cross-flow ultrafiltration of BSA and dextran solutions. J Membr Sci 90(1–2):183–189. doi:10.1016/0376-7388(94)80045-6

Cui ZF, Wright KIT (1996) Flux enhancements with gas sparging in downwards crossflow ultrafiltration: Performance and mechanism. J Membr Sci 117(1–2):109–116. doi:10.1016/0376-7388(96)00040-3

Cui ZF, Chang S, Fane AG (2003) The use of gas bubbling to enhance membrane processes. J Membr Sci 221(1–2):1–35. doi:10.1016/S0376-7388(03)00246-1

Cui X, Choo KH (2013) Granular iron oxide adsorbents to control natural organic matter and membrane fouling in ultrafiltration water treatment. Water Res 47(13):4227–4237

Czekaj P, López F, Güell C (2000) Membrane fouling during microfiltration of fermented beverages. J Membr Sci 166(2):199–212

De Beer D, Stoodley P (2013) Microbial Biofilms, in The Prokaryotes: Applied bacteriology and biotechnology Rosenberg E, et al (eds) Springer Berlin Heidelberg: Berlin, Heidelberg. p. 343–372

De La Torre T, Lesjean B, Drews A, Kraume M (2008) Monitoring of transparent exopolymer particles (TEP) in a membrane bioreactor (MBR) and correlation with other fouling indicators. Water Sci Technol 58:1903–1909

Diagne F, et al (2012) Polyelectrolyte and silver nanoparticle modification of microfiltration membranes to mitigate organic and bacterial fouling. Environ Sci Technol 46(7):4025–4033

Drews A, Lee CH, Kraume M (2006a) Membrane fouling—a review on the role of EPS. Desalination 200(1–3):186–188

Drews A, Vocks M, Iversen V, Lesjean B, Kraume M (2006b) Influence of unsteady membrane bioreactor operation on EPS formation and filtration resistance. Desalination 192(1–3):1–9

Drews A, Mante J, Iversen V, Vocks M, Lesjean B, Kraume M (2007) Impact of ambient conditions on SMP elimination and rejection in MBRs. Water Res 41(17):3850–3858

Drews A, Prieske H, Meyer EL, Senger G, Kraume M (2010) Advantageous and detrimental effects of air sparging in membrane filtration: bubble movement, exerted shear and particle classification. Desalination 250(3):1083–1086. doi:10.1016/j.desal.2009.09.113

Ducom G, Matamoros H, Cabassud C (2002) Air sparging for flux enhancement in nanofiltration membranes: application to O/W stabilised and non-stabilised emulsions. J Membr Sci 204(1–2):221–236. doi:10.1016/S0376-7388(02)00044-3

Escobar I (2005) Hydraulic and chemical cleaning of cellulose acetate ultrafiltration membranes. In: AIChE annual meeting, conference proceedings, p 2031

Essemiani K, Ducom G, Cabassud C, Liné A (2001) Spherical cap bubbles in a flat sheet nanofiltration module: experiments and numerical simulation. Chem Eng Sci 56(21–22):6321–6327. doi:10.1016/S0009-2509(01)00282-2

Evenblij H, van der Graaf JHJM (2004) Occurrence of EPS in activated sludge from a membrane bioreactor treating municipal wastewater. Water Sci Technol 50:293–300

Fan X, Urbain V, Qian Y, Manem J (2000) Ultrafiltration of activated sludge with ceramic membranes in a cross-flow membrane bioreactor process. Water Sci Technol 41(10/11):243–250

Fan F, Zhou H, Husain H (2006) Identification of wastewater sludge characteristics to predict critical flux for membrane bioreactor processes. Water Res 40(2):205–212

Fane AG (2005) Towards sustainability in membrane processes for water and wastewater processing. In: Proceedings of the international congress on membranes and membrane processes (ICOM), Seoul, Korea

Field RW, Wu D, Howell JA, Gupta BB (1995) Critical flux concept for microfiltration fouling. J Membr Sci 100(3):259–272

Finnigan SM, Howell JA (1989) Effect of pulsatile flow on ultrafiltration fluxes in a baffled tubular membrane system. Chem Eng Res Des 67(3):278–282

Flemming HC (1997) Reverse osmosis membrane biofouling. Exp Thermal Fluid Sci 14(4):382–391

Flemming HC, Schaule G (1988) Biofouling on membranes—a microbiological approach. Desalination 70(1–3):95–119. doi:10.1016/0011-9164(88)85047-1

Flemming HC, Schaule G, Griebe T, Schmitt J, Tamachkiarowa A (1997) Biofouling—the Achilles heel of membrane processes. Desalination 113(2–3):215–225

Geng Z, Hall ER (2007) A comparative study of fouling-related properties of sludge from conventional and membrane enhanced biological phosphorus removal processes. Water Res 41 (19):4329–4338

Genkin G, Waite TD, Fane T, Chang S (2005) The effet of axial vibratins on the filtration performance of submerged hollow fibre membranes. In: Proceedings of the international congress on membranes and membrane processes (ICOM), Seoul, Korea

Ghaffour N, Jassim R, Khir T (2004) Flux enhancement by using helical baffles in ultrafiltration of suspended solids. Desalination 167(1–3):201–207. doi:10.1016/j.desal.2004.06.129

Ghosh R (2006) Enhancement of membrane permeability by gas-sparging in submerged hollow fibre ultrafiltration of macromolecular solutions: role of module design. J Membr Sci 274(1–2):73–82. doi:10.1016/j.memsci.2005.08.002

Gil JA, Túa L, Rueda A, Montaño B, Rodríguez M, Prats D (2010) Monitoring and analysis of the energy cost of an MBR. Desalination 250(3):997–1001. doi:10.1016/j.desal.2009.09.089

Guglielmi G, Chiarani D, Judd SJ, Andreottola G (2007a) Flux criticality and sustainability in a hollow fibre submerged membrane bioreactor for municipal wastewater treatment. J Membr Sci 289(1–2):241–248

Guglielmi G, Saroj DP, Chiarani D, Andreottola G (2007b) Sub-critical fouling in a membrane bioreactor for municipal wastewater treatment: experimental investigation and mathematical modelling. Water Res 41(17):3903–3914

Guo WS, Vigneswaran S, Ngo HH (2004) A rational approach in controlling membrane fouling problems: pretreatments to a submerged hollow fiber membrane system. In: Proceedings of the water environment-membrane technology conference, Seoul, Korea

Guo WS, Vigneswaran S, Ngo HH, Xing W (2007) Experimental investigation on acclimatized wastewater for membrane bioreactors. Desalination 207(1–3):383–391. doi:10.1016/j.desal. 2006.07.013

Guo W, Ngo HH, Li J (2012) A mini-review on membrane fouling. Bioresour Technol 122:27–34

Gupta BB, Blanpain P, Jaffrin MY (1992) Permeate flux enhancement by pressure and flow pulsations in microfiltration with mineral membranes. J Membr Sci 70(2–3):257–266. doi:10. 1016/0376-7388(92)80111-V

Gupta BB, Zaboubi B, Jaffrin MY (1993) Scaling up pulsatile filtration flow methods to a pilot apparatus equipped with mineral membranes. J Membr Sci 80(1):13–20. doi:10.1016/0376-7388(93)85128-J

Hasson D, Drak A, Semiat R (2001) Inception of CaSO$_4$ scaling on RO membranes at various water recovery levels. Desalination 139(1–3):73–81

Hilal N, Ogunbiyi OO, Miles NJ, Nigmatullin R (2005) Methods employed for control of fouling in MF and UF membranes: a comprehensive review. Sep Sci Technol 40(10):1957–2005

Holbrook RD, Higgins MJ, Murthy SN, Fonseca AD, Fleischer EJ, Daigger GT, Grizzard TJ, Love NG, Novak JT (2004) Effect of alum addition on the performance of submerged membranes for wastewater treatment. Water Environ Res 76(7):2699–2702

Holdich RG, Zhang GM (1992) Crossflow microfiltration incorporating rotational fluid flow. Chem Eng Res Des 70(A5):527–536

Hong S, Elimelech M (1997) Chemical and physical aspects of natural organic matter (NOM) fouling of nanofiltration membranes. J Membrane Sci 132(2):159–181

Hong SP, Bae TH, Tak TM, Hong S, Randall A (2002) Fouling control in activated sludge submerged hollow fiber membrane bioreactors. Desalination 143(3):219–228

Houari A, Seyer D, Couquard F, Kecili K, Démocrate C, Heim V, Di Martino P (2010) Characterization of the biofouling and cleaning efficiency of nanofiltration membranes. Biofouling 26(1):15–21

Howell JA, Field RW, Wu D (1993) Yeast cell microfiltration: flux enhancement in baffled and pulsatile flow systems. J Membr Sci 80(1):59–71. doi:10.1016/0376-7388(93)85132-G

Howell JA, Chua HC, Arnot TC (2004) In situ manipulation of critical flux in a submerged membrane bioreactor using variable aeration rates, and effects of membrane history. J Membr Sci 242(1–2):13–19. doi:10.1016/j.memsci.2004.05.013

Huang X, Liu R, Qian Y (2000) Behaviour of soluble microbial products in a membrane bioreactor. Process Biochem 36(5):401–406

Huang X, Xiao K, Shen Y (2010) Recent advances in membrane bioreactor technology for wastewater treatment in China. Front Environ Sci Eng China 4(3):245–271

Huang J, Arthanareeswaran G, Zhang K (2012a) Effect of silver loaded sodium zirconium phosphate (nanoAgZ) nanoparticles incorporation on PES membrane performance. Desalination 285:100–107

Huang J, Zhang K, Wang K, Xie Z, Ladewig B, Wang H (2012b) Fabrication of polyethersulfone-mesoporous silica nanocomposite ultrafiltration membranes with antifouling properties. J Membr Sci 423–424:362–370

Huang J, Wang H, Zhang K (2014) Modification of PES membrane with Ag-SiO2: reduction of biofouling and improvement of filtration performance. Desalination 336(1):8–17

Huyskens C, Brauns E, Van Hoof E, De Wever H (2008) A new method for the evaluation of the reversible and irreversible fouling propensity of MBR mixed liquor. J Membr Sci 323(1): 185–192

Hwang BK, Lee WN, Park PK, Lee CH, Chang IS (2007) Effect of membrane fouling reducer on cake structure and membrane permeability in membrane bioreactor. J Membr Sci 288(1–2):149–156. doi:10.1016/j.memsci.2006.11.032

Imasaka T, So H, Matsushita K, Furukawa T, Kanekuni N (1993) Application of gas-liquid two-phase cross-flow filtration to pilot-scale methane fermentation. Drying Technol 11(4): 769–785

Itonaga T, Kimura K, Watanabe Y (2004) Influence of suspension viscosity and colloidal particles on permeability of membrane used in membrane bioreactor (MBR). Water Sci Technol 50:301–309

Jang N, Ren X, Cho J, Kim IS (2006) Steady-state modeling of bio-fouling potentials with respect to the biological kinetics in the submerged membrane bioreactor (SMBR). J Membr Sci 284(1–2):352–360

Jeong TY, Cha GC, Yoo IK, Kim DJ (2007) Characteristics of bio-fouling in a submerged MBR. Desalination 207(1–3):107–113. doi:10.1016/j.desal.2006.07.006

Ji J, Qiu J, Fs Wong, Li Y (2008) Enhancement of filterability in MBR achieved by improvement of supernatant and floc characteristics via filter aids addition. Water Res 42(14):3611–3622. doi:10.1016/j.watres.2008.05.022

Ji J, Qiu J, Wai N, Wong FS, Li Y (2010) Influence of organic and inorganic flocculants on physical-chemical properties of biomass and membrane-fouling rate. Water Res 44(5):1627–1635. doi:10.1016/j.watres.2009.11.013

Jiang T, Kennedy MD, Guinzbourg BF, Vanrolleghem PA, Schippers JC (2005) Optimising the operation of a MBR pilot plant by quantitative analysis of the membrane fouling mechanism. Water Sci Technol 51:19–25

Judd S (2002) Submerged membrane bioreactors: flat plate or hollow fibre? Filtr Sep 39(5):30–31. doi:10.1016/S0015-1882(02)80169-0

Judd S (2006) Principles and applications of membrane bioreactors in water and waste water treatment, 1st edn. Elsevier, UK

Judd S (2007) The status of membrane bioreactor technology. Trends Biotechnol 26(2):109–116

Judd S (2011) Chapter 3—Design, operation and maintenance. In: The MBR Book, 2nd edn. Butterworth-Heinemann, Oxford, pp 209–288. doi:http://dx.doi.org/10.1016/B978-0-08-096682-3.10003-4

Judd S, Jefferson B (2003) Membranes for industrial wastewater recovery and re-use. Elesevier

Judd S, Alvarez-vazquez H, Jefferson B (2006) The impact of intermittent aeration on the operation of air-lift tubular membrane bioreactors under sub-critical conditions. Sep Sci Technol 41(7):1293–1302. doi:10.1080/01496390600634541

Kang IJ, Yoon SH, Lee CH (2002) Comparison of the filtration characteristics of organic and inorganic membranes in a membrane-coupled anaerobic bioreactor. Water Res 36(7):1803–1813

Kang IJ, Lee CH, Kim KJ (2003) Characteristics of microfiltration membranes in a membrane coupled sequencing batch reactor system. Water Res 37(5):1192–1197

Kang ST, Subramani A, Hoek EMV, Deshusses MA, Matsumoto MR (2004) Direct observation of biofouling in cross-flow microfiltration: mechanisms of deposition and release. J Membr Sci 244(1–2):151–165. doi:10.1016/j.memsci.2004.07.011

Kappachery S, Paul D, Yoon J, Kweon JH (2010) Vanillin, a potential agent to prevent biofouling of reverse osmosis membrane. Biofouling 26(6):667–672

Khan MMT, Stewart PS, Moll DJ, Mickols WE, Nelson SE, Camper AK (2011) Characterization and effect of biofouling on polyamide reverse osmosis and nanofiltration membrane surfaces. Biofouling 27(2):173–183

Kim IS, Jang N (2006) The effect of calcium on the membrane biofouling in the membrane bioreactor (MBR). Water Res 40(14):2756–2764

Kim J, Jang M, Chio H, Kim S (2004) Characteristics of membrane and module affecting membrane fouling. In: Proceedings of the water environment-membrane technology conference, Seoul, Korea

Kim J, Shi W, Yuan Y, Benjamin MM (2007) A serial filtration investigation of membrane fouling by natural organic matter. J Membr Sci 294(1–2):115–126

Kimura K, Hane Y, Watanabe Y, Amy C, Ohkuma N (2004) Irreversible membrane fouling during filtration of surface water. Water Res 38:3431–3441

Kimura K, Yamato N, Yamamura H, Watanabe Y (2005) Membrane fouling in pilot-scale membrane bioreactors (MBRs) treating municipal wastewater. Environ Sci Technol 39 (16):6293–6299

Kochkodan V, Johnson DJ, Hilal N (2014) Polymeric membranes: Surface modification for minimizing (bio) colloidal fouling. Adv Colloid Interfac 206:116—140

Kochkodan V, Hilal N (2015) A comprehensive review on surface modified polymer membranes for biofouling mitigation. Desalination 356:187–207. doi:10.1016/j.desal.2014.09.015

Koga H, Kitaoka T, Wariishi H (2009) In situ synthesis of silver nanoparticles on zinc oxide whiskers incorporated in a paper matrix for antibacterial applications. J Mater Chem 19 (15):2135–2140

Komlenic R (2010) Rethinking the causes of membrane biofouling. Filtr Sep 47(5):26–28. doi:10. 1016/S0015-1882(10)70211-1

Koseoglu H, Yigit NO, Iversen V, Drews A, Kitis M, Lesjean B, Kraume M (2008) Effects of several different flux enhancing chemicals on filterability and fouling reduction of membrane bioreactor (MBR) mixed liquors. J Membr Sci 320(1–2):57–64. doi:10.1016/j.memsci.2008. 03.053

Kox LSDM (2004) Membrane bioreactor in Varsseveld, the Dutch approach. In: Proceedings of the water environment-membrane technology conference, Seoul, Korea

Kramer JF, Tracey DA (1995) The solution to reverse osmosis biofouling. In Proceedings of IDA world congress on desalination and water use, Abu Dhabi, Saudi Arabia, November 19[25]95; 4:33–44

Kraume M, Wedi D, Schaller J, Iversen V, Drews A (2009) Fouling in MBR—what use are lab investigations for full scale operation. Desalination 236:94–103

Laabs CN, Amy GL, Jekel M (2006) Understanding the size and character of fouling-causing substances from effluent organic matter (EfOM) in low-pressure membrane filtration. Environ Sci Technol 40(14):4495–4499. doi:10.1021/es060070r

Laborie S, Cabassud C, Durand-Bourlier L, Lainé JM (1998) Fouling control by air sparging inside hollow fibre membranes—effects on energy consumption. Desalination 118(1–3):189–196. doi:10.1016/S0011-9164(98)00124-6

Le Clech P, Jefferson B, Chang IS, Judd SJ (2003) Critical flux determination by the flux-step method in a submerged membrane bioreactor. J Membr Sci 227(1–2):81–93

Lebegue J, Heran M, Grasmick A (2008) Membrane bioreactor: distribution of critical flux throughout an immersed HF bundle. Desalination 231(1–3):245–252

Le-Clech P, Fane A, Leslie G, Childress A (2005a) The operator's perspective. Filtr Sep 42:20–23

Le-Clech P, Jefferson B, Judd SJ (2005b) A comparison of submerged and sidestream tubular membrane bioreactor configurations. Desalination 173(2):113–122

Le-Clech P, Cao Z, Wan PY, Wiley DE, Fane AG (2006a) The application of constant temperature anemometry to membrane processes. J Membr Sci 284(1–2):416–423

Le-Clech P, Chen V, Fane TAG (2006b) Fouling in membrane bioreactors used in wastewater treatment. J Membr Sci 284(1–2):17–53

Le-Clech P, Marselina Y, Stuetz R, Chen V (2006c) Fouling visualisation of soluble microbial product models in MBRs. Desalination 199(1–3):477–479

Lee S, Lee CH (2000) Effect of operating conditions on $CaSO_4$ scale formation mechanism in nanofiltration for water softening. Water Res 34(15):3854–3866

Lee CK, Chang WG, Ju YH (1993) Air slugs entrapped cross-flow filtration of bacterial suspensions. Biotechnol Bioeng 41(5):525–530. doi:10.1002/bit.260410504

Lee S, Kim J, Lee CH (1999) Analysis of $CaSO_4$ scale formation mechanism in various nanofiltration modules. J Membr Sci 163(1):63–74

Lee J, Ahn WY, Lee CH (2001a) Comparison of the filtration characteristics between attached and suspended growth microorganisms in submerged membrane bioreactor. Water Res 35 (10):2435–2445

Lee JC, Kim JS, Kang IJ, Cho MH, Park PK, Lee CH (2001b) Potential and limitations of alum or zeolite addition to improve the performance of a submerged membrane bioreactor. Water Sci Technol 43:59–66

Lee W, Kang S, Shin H (2003) Sludge characteristics and their contribution to microfiltration in submerged membrane bioreactors. J Membr Sci 216(1–2):217–227

Lee DH, Kim H-I, Kim SS (2004) Surface modification of polymeric membranes by UV grafting. In: Advanced materials for membrane separations. ACS Symposium Series, American Chemical Society 876:281-299

Lee SY, Kim HJ, Patel R, Im SJ, Kim JH, Min BR (2007) Silver nanoparticles immobilized on thin film composite polyamide membrane: characterization, nanofiltration, antifouling properties. Polym Adv Technol 18(7):562–568

Leiknes TO (2012) Membrane bioreactors. In: Peinemann K-V, Nunes SP (eds) Membranes for water treatment. Wiley, New York, pp 193–226

Lesjean B, Rosenberger S, Laabs C, Jekel M, Gnirss R, Amy G (2005) Correlation between membrane fouling and soluble/colloidal organic substances in membrane bioreactors for municipal wastewater treatment. Water Sci Technol 51:1–8

Li XY, Yang SF (2007) Influence of loosely bound extracellular polymeric substances (EPS) on the flocculation, sedimentation and dewaterability of activated sludge. Water Res 41(5):1022–1030

Li QY, Cui ZF, Pepper DS (1997) Effect of bubble size and frequency on the permeate flux of gas sparged ultrafiltration with tubular membranes. Chem Eng J 67(1):71–75. doi:10.1016/S1385-8947(97)00016-8

Li QY, Ghosh R, Bellara SR, Cui ZF, Pepper DS (1998) Enhancement of ultrafiltration by gas sparging with flat sheet membrane modules. Sep Purif Technol 14(1–3):79–83. doi:10.1016/S1383-5866(98)00062-8

Li M, Zhao Y, Zhou S, Xing W, Wong FS (2007) Resistance analysis for ceramic membrane microfiltration of raw soy sauce. J Membr Sci 299(1–2):122–129. doi:10.1016/j.memsci.2007.04.033

Liang S, Liu C, Song L (2007) Soluble microbial products in membrane bioreactor operation: behaviors, characteristics, and fouling potential. Water Res 41(1):95–101

Liao BQ, Bagley DM, Kraemer HE, Leppard GG, Liss SN (2004) A review of biofouling and its control in membrane separation bioreactors. Water Environ Res 76(5):425–436

Lim BR, Ahn KH, Song KG, Woo CJ (2004) Microbial community in biofilm on membrane surface of submerged MBR: effect of in-line cleaning chemical agent. In: Proceedings of the water environment-membrane technology conference, Seoul, Korea

Lin CJ, Shirazi S, Rao P (2005) Mechanistic model for $CaSO_4$ fouling on nanofiltration membrane. J Environ Eng 131(10):1387–1392

Lin X, et al (2015) Composite ultrafiltration membranes from polymer and its quaternary phosphonium-functionalized derivative with enhanced water flux. J Membrane Sci 482: p. 67–75.

Liu R, Huang X, Xi J, Qian Y (2005) Microbial behaviour in a membrane bioreactor with complete sludge retention. Process Biochem 40(10):3165–3170

Liu Y, Liu H, Cui L, Zhang K (2012) The ratio of food-to-microorganism (F/M) on membrane fouling of anaerobic membrane bioreactors treating low-strength wastewater. Desalination 297:97–103

Lyko S, Al-Halbouni D, Wintgens T, Janot A, Hollender J, Dott W, Melin T (2007) Polymeric compounds in activated sludge supernatant—characterisation and retention mechanisms at a full-scale municipal membrane bioreactor. Water Res 41(17):3894–3902

Ma H, Hakim LF, Bowman CN, Davis RH (2001a) Factors affecting membrane fouling reduction by surface modification and backpulsing. J Membr Sci 189(2):255–270

Ma H, Nielsen DR, Bowman CN, Davis RH (2001b) Membrane surface modification and backpulsing for wastewater treatment. Sep Sci Technol 36(7):1557–1573

Ma W, Zhang J, Wang X, Wang S (2007) Effect of PMMA on crystallization behavior and hydrophilicity of poly(vinylidene fluoride)/poly(methyl methacrylate) blend prepared in semi-dilute solutions. Appl Surf Sci 253(20):8377–8388

Mackay ME, Mackley MR, Wang Y (1991) Oscillatory flow within tubes containing wall or central baffles. Chem Eng Res Des 69(6):506–513

Mansouri J, Harrisson S, Chen V (2010) Strategies for controlling biofouling in membrane filtration systems: challenges and opportunities. J Mater Chem 20(22):4567–4586

Martinelli L, Guigui C, Line A (2010) Characterisation of hydrodynamics induced by air injection related to membrane fouling behaviour. Desalination 250(2):587–591. doi:10.1016/j.desal.2009.09.029

Matsumura Y, Yoshikata K, Kunisaki SI, Tsuchido T (2003) Mode of bactericidal action of silver zeolite and its comparison with that of silver nitrate. Appl Environ Microbiol 69(7):4278–4281

Mayer M, Braun R, Fuchs W (2006) Comparison of various aeration devices for air sparging in crossflow membrane filtration. J Membr Sci 277(1–2):258–269. doi:10.1016/j.memsci.2005.10.035

Meng F, Yang F (2007) Fouling mechanisms of deflocculated sludge, normal sludge, and bulking sludge in membrane bioreactor. J Membr Sci 305(1–2):48–56. doi:10.1016/j.memsci.2007.07.038

Meng F, Shi B, Yang F, Zhang H (2007) Effect of hydraulic retention time on membrane fouling and biomass characteristics in submerged membrane bioreactors. Bioprocess Biosyst Eng 30:359–367

Meng F, et al. (2007) Characterization of cake layer in submerged membrane bioreactor. Environ Sci Technol 41(11): p. 4065–4070

Meng F, Chae SR, Drews A, Kraume M, Shin HS, Yang F (2009) Recent advances in membrane bioreactors (MBRs): Membrane fouling and membrane material. Water Res 43(6):1489–1512

Mercier M, Fonade C, Lafforgue-Delorme C (1997) How slug flow can enhance the ultrafiltration flux in mineral tubular membranes. J Membr Sci 128(1):103–113. doi:10.1016/S0376-7388(96)00317-1

Metzger U, Le-Clech P, Stuetz RM, Frimmel FH, Chen V (2007) Characterisation of polymeric fouling in membrane bioreactors and the effect of different filtration modes. J Membr Sci 301(1–2):180–189

Miura Y, Watanabe Y, Okabe S (2007) Membrane biofouling in pilot-scale membrane bioreactors (MBRs) treating municipal wastewater: Impact of biofilm formation. Environ Sci Technol 41(2):632–638. doi:10.1021/es0615371

Miyashita S, Honjyo K, Kato O, Watari K, Takashima T, Itakura M, Okazaki H, Kinoshita I, Inoue N (2000) Gas diffuser for aeration vessel of membrane assembly. United States Patent Patent

Mo Y, et al (2012) Improved antifouling properties of polyamide nanofiltration membranes by reducing the density of surface carboxyl groups. Environ Sci Technol. 46(24):13253–13261

Nagaoka H, Nemoto H (2005) Influence of extracellular polymeric substance on nitrogen removal in an intermittently-aerated membrane bioreactor. Water Sci Technol 51:151–158

Nagaoka H, Ueda S, Miya A (1996) Influence of bacterial extracellular polymers on the membrane separation activated sludge process. Water Sci Technol 34:165–172

Ndinisa NV, Fane AG, Wiley DE, Fletcher DF (2006) Fouling control in a submerged flat sheet membrane system: part II—two-phase flow characterization and CFD simulations. Sep Sci Technol 41(7):1411–1445. doi:10.1080/01496390600633915

Neu TR, Marshall KC (1990) Bacterial polymers: physicochemical aspects of their interactions at interfaces. J Biomater Appl 5(2):107–133

Ng CA, Sun D, Zhang J, Chua HC, Bing W, Tay S, Fane A (2005) Strategies to improve the sustainable operation of membrane bioreactors. In: Proceedings of the international desalination association conference, Singapore

Ng HY, Tan TW, Ong SL (2006) Membrane fouling of submerged membrane bioreactors: impact of mean cell residence time and the contributing factors. Environ Sci Technol 40(8):2706–2713

Nilson JA, DiGiano FA (1996) Influence of NOM composition on nanofiltration. J Am Water Works Assoc 88(5):53–66

Ognier S, Wisniewski C, Grasmick A (2002a) Characterisation and modelling of fouling in membrane bioreactors. Desalination 146(1–3):141–147

Ognier S, Wisniewski C, Grasmick A (2002b) Influence of macromolecule adsorption during filtration of a membrane bioreactor mixed liquor suspension. J Membr Sci 209(1):27–37

Ognier S, Wisniewski C, Grasmick A (2002c) Membrane fouling during constant flux filtration in membrane bioreactors. Membr Technol 147:6–10

Ognier S, Wisniewski C, Grasmick A (2004) Membrane bioreactor fouling in sub-critical filtration conditions: a local critical flux concept. J Membr Sci 229(1–2):171–177

Oh SJ, Kim N, Lee YT (2009) Preparation and characterization of PVDF/TiO2 organic-inorganic composite membranes for fouling resistance improvement. J Membr Sci 345(1–2):13–20

Okazaki M, Kimura S (1984) scale formation on reverse-osmosis membranes. J Chem Eng Jpn 17(2):145–151

Pan JR, Su YC, Huang C, Lee HC (2010) Effect of sludge characteristics on membrane fouling in membrane bioreactors. J Membr Sci 349(1–2):287–294

Pang CM, Hong P, Guo H, Liu WT (2005) Biofilm formation characteristics of bacterial isolates retrieved from a reverse osmosis membrane. Environ Sci Technol 39(19):7541–7550

Park D, Lee DS, Park JM (2005) Continuous biological ferrous iron oxidation in a submerged membrane bioreactor. Water Sci Technol 51:59–68

Pervov AG (1991) Scale formation prognosis and cleaning procedure schedules in reverse osmosis systems operation. Desalination 83(1-3):77–118

Peuchot MM, Aim RB (1992) Improvement of cross-flow microfiltration performances with flocculation. J Membr Sci 68:241–248

Phattaranawik J, Fane AG, Pasquier ACS, Wu B (2007) Membrane bioreactor with bubble-size transformer: design and fouling control. AIChE J 53(1):243–248. doi:10.1002/aic.11040

Phuntsho S, et al (2012) Influence of temperature and temperature difference in the performance of forward osmosis desalination process. J Membrane Sci 415–416:734–744

Pollice A, Brookes A, Jefferson B, Judd S (2005) Sub-critical flux fouling in membrane bioreactors —a review of recent literature. Desalination 174(3):221–230

Porcelli N, Judd S (2010) Chemical cleaning of potable water membranes: a review. Sep Purif Technol 71(2):137–143

Potts DE, Ahlert RC, Wang SS (1981) A critical review of fouling of reverse osmosis membranes. Desalination 36(3):235–264

Psoch C, Schiewer S (2005) Long-term study of an intermittent air sparged MBR for synthetic wastewater treatment. J Membr Sci 260(1–2):56–65. doi:10.1016/j.memsci.2005.03.021

Psoch C, Schiewer S (2006) Anti-fouling application of air sparging and backflushing for MBR. J Membr Sci 283(1–2):273–280. doi:10.1016/j.memsci.2006.06.042

Rabie HR, Cote P, Singh M, Janson A (2003) Cyclic aeration system for submerged membrane modules. United States Patent Patent

Ramesh A, Lee DJ, Wang ML, Hsu JP, Juang RS, Hwang KJ, Liu JC, Tseng SJ (2006) Biofouling in membrane bioreactor. Sep Sci Technol 41(7):1345–1370

Ramesh A, Lee DJ, Lai JV (2007) Membrane biofouling by extracellular polymeric substances or soluble mcirobial products from membrane bioreactor sludge. Appl Microbiol Biotechnol 74:699–707

Rana D, Matsuura T (2010) Surface modifications for antifouling membranes. Chem Rev 110 (4):2448–2471

Ridgway HF, Rigby MG, Argo DG (1985) Bacterial adhesion and fouling of reverse osmosis membranes. J Am Water Works Assoc 77(7):97–106

Rosenberger S, Kraume M (2002) Filterability of activated sludge in membrane bioreactors. Desalination 146(1–3):373–379

Rosenberger S, Krüger U, Witzig R, Manz W, Szewzyk U, Kraume M (2002a) Performance of a bioreactor with submerged membranes for aerobic treatment of municipal waste water. Water Res 36(2):413–420

Rosenberger S, Kubin K, Kraume M (2002b) Rheology of activated sludge in membrane activation reactors. Chemie-Ingenieur-Technik 74(4):487–494 + 373

Roudman AR, Digiano FA (2000) Surface energy of experimental and commercial nanofiltration membranes: Effects of wetting and natural organic matter fouling. J Membr Sci 175(1):61–73

Samir K, Kusakabe K, Raghunathan K, Fan LS (1992) Mechanism of heat transfer in bubbly liquid and liquid-solid systems: single bubble injection. AIChE J 38(5):733–741

Schafer A, Fane A, Waite T (2005) Nanofiltation: principles and applications, 1st edn. Elsevier Advanced Technology, Oxford, UK

Schneider RP, Ferreira LM, Binder P, Bejarano EM, Góes KP, Slongo E, Machado CR, Rosa GMZ (2005) Dynamics of organic carbon and of bacterial populations in a conventional pretreatment train of a reverse osmosis unit experiencing severe biofouling. J Membr Sci 266 (1–2):18–29. doi:10.1016/j.memsci.2005.05.006

Schoeberl P, Brik M, Bertoni M, Braun R, Fuchs W (2005) Optimization of operational parameters for a submerged membrane bioreactor treating dyehouse wastewater. Sep Purif Technol 44 (1):61–68. doi:10.1016/j.seppur.2004.12.004

Serra C, Durand-Bourlier L, Clifton MJ, Moulin P, Rouch JC, Aptel P (1999) Use of air sparging to improve backwash efficiency in hollow-fiber modules. J Membr Sci 161(1–2):95–113. doi:10.1016/S0376-7388(99)00106-4

Sheikholaslami R (1999) Composite fouling—inorganic and biological: a review. Environ Prog 18 (2):113–122

Sheikholeslami R (1999) Fouling mitigation in membrane processes. Desalination 123(1):45–53

Shin HS, Kang ST (2003) Characteristics and fates of soluble microbial products in ceramic membrane bioreactor at various sludge retention times. Water Res 37(1):121–127

Shirazi S, Lin CJ, Doshi S, Agarwal S, Rao P (2006) Comparison of fouling mechanism by $CaSO_4$ and $CaHPO_4$ on nanofiltration membranes. Sep Sci Technol 41(13):2861–2882. doi:10.1080/01496390600854529

Shirazi S, Lin CJ, Chen D (2010) Inorganic fouling of pressure-driven membrane processes—a critical review. Desalination 250(1):236–248

Simon A, Price WE, Nghiem LD (2013) Influence of formulated chemical cleaning reagents on the surface properties and separation efficiency of nanofiltrationmembranes. J Membrane Sci 432:73–82

Smith PJ, Vigneswaran S, Ngo HH, Ben-Aim R, Nguyen H (2005) Design of a generic control system for optimising back flush durations in a submerged membrane hybrid reactor. J Membr Sci 255(1–2):99–106. doi:10.1016/j.memsci.2005.01.026

Sofia A, Ng WJ, Ong SL (2004) Engineering design approaches for minimum fouling in submerged MBR. Desalination 160(1):67–74. doi:10.1016/S0011-9164(04)90018-5

Sperandio M, Masse A, Espinosa-Bouchot MC, Cabassud C (2005) Characterization of sludge structure and activity in submerged membrane bioreactor. Water Sci Technol 52(10–11):401–408

Speth T, Summers R, Gusses A (1998) Nanofiltration foulants from a treated surface water. Environ Sci Technol 32:3612–3617

Spiazzi E, Lenoir J, Grangeon A (1993) A new generator of unsteady-state flow regime in tubular membranes as an anti-fouling technique: a hydrodynamic approach. J Membr Sci 80(1):49–57. doi:10.1016/0376-7388(93)85131-F

Sun Y, Huang X, Chen F, Wen XA (2004) dual functional filtration/aeration membrane bioreactor for domestic wastewater treatment. In: Proceedings of the water environment-membrane technology conference, Seoul, Korea

Sun Y, Wang Y, Huang X (2007) Relationship between sludge settleability and membrane fouling in a membrane bioreactor. Front Environ Sci Eng China 1(2):221–225

Taha T, Cui ZF (2002) CFD modelling of gas-sparged ultrafiltration in tubular membranes. J Membr Sci 210(1):13–27. doi:10.1016/S0376-7388(02)00360-5

Tansel B, Sager J, Garland J, Xu S, Levine L, Bisbee P (2006) Deposition of extracellular polymeric substances (EPS) and microtopographical changes on membrane surfaces during intermittent filtration conditions. J Membr Sci 285(1–2):225–231. doi:10.1016/j.memsci.2006.08.031

Tao G, Kekre K, Wei Z, Lee TC, Viswanath B, Seah H (2005) Membrane bioreactors for water reclamation. Water Sci Technol 51:431–440

Tardieu E, Grasmick A, Geaugey V, Manem J (1998) Hydrodynamic control of bioparticle deposition in a MBR applied to wastewater treatment. J Membr Sci 147(1):1–12. doi:10.1016/S0376-7388(98)00091-X

Teychene B, Guigui C, Cabassud C (2011) Engineering of an MBR supernatant fouling layer by fine particles addition: a possible way to control cake compressibility. Water Res 45(5):2060–2072. doi:10.1016/j.watres.2010.12.018

Tiraferri A, et al (2012) Superhydrophilic thin-film composite forward osmosis membranes for organic fouling control: Fouling behavior and antifouling mechanisms. Environ Sci Technol 46(20):11135–11144

Tradieu E, Grasmick A, Geaugey V, Manem J (1998) Hydrodynamic control of bioparticle deposition in a MBR applied to waste water treatment. Membr Sci 147:1–12

Trussell RS, Merlo RP, Hermanowicz SW, Jenkins D (2006) The effect of organic loading on process performance and membrane fouling in a submerged membrane bioreactor treating municipal wastewater. Water Res 40(14):2675–2683. doi:10.1016/j.watres.2006.04.020

Trussell RS, Merlo RP, Hermanowicz SW, Jenkins D (2007) Influence of mixed liquor properties and aeration intensity on membrane fouling in a submerged membrane bioreactor at high mixed liquor suspended solids concentrations. Water Res 41(5):947–958. doi:10.1016/j.watres.2006.11.012

Ueda T, Hata K, Kikuoka Y (1996) Treatment of domestic sewage from rural settlements by a membrane bioreactor. Water Sci Technol 34:189–196

Vallero MVG, Lettinga G, Lens PNL (2005) High rate sulfate reduction in a submerged anaerobic membrane bioreactor (SAMBaR) at high salinity. J Membr Sci 253(1–2):217–232. doi:10.1016/j.memsci.2004.12.032

Van De Lisdonk CAC, Van Paassen JAM, Schippers JC (2000) Monitoring scaling in nanofiltration and reverse osmosis membrane systems. Desalination 132(1–3):101–108

Verbeck J, Worm G, Futselaar H, Dijk JCV (2000) Combined air water flush in dead-end ultrafiltration. In: Proceedings of the IWA conference on drinking and industrial water production, Paris, pp 655–663

Verrecht B, Judd S, Guglielmi G, Brepols C, Mulder JW (2008) An aeration energy model for an immersed membrane bioreactor. Water Res 42(19):4761–4770. doi:10.1016/j.watres.2008.09.013

Visvanathan C, Ben aïm R (1989) Studies on colloidal membrane fouling mechanisms in crossflow microfiltration. J Membr Sci 45(1–2):3–15

Visvanathan C, Yang BS, Muttamara S, Maythanukhraw R (1997) Application of air backflushing technique in membrane bioreactor. Water Sci Technol 36(12):259–266. doi:10.1016/S0273-1223(97)00727-0

Vrouwenvelder JS, van der Kooij D (2003) Diagnosis of fouling problems of NF and RO membrane installations by a quick scan. Desalination 153(1–3):121–124

Vrouwenvelder JS, van Loosdrecht MCM, Kruithof JC (2011) Early warning of biofouling in spiral wound nanofiltration and reverse osmosis membranes. Desalination 265(1–3):206–212

Wakeman R, Williams C (2002a) Additional techniques to improve microfiltration. Sep Purif Technol 26:3–18

Wakeman RJ, Williams CJ (2002b) Additional techniques to improve microfiltration. Sep Purif Technol 26(1):3–18. doi:10.1016/S1383-5866(01)00112-5

Wang Y, Howell JA, Field RW, Wu D (1994) Simulation of cross-flow filtration for baffled tubular channels and pulsatile flow. J Membr Sci 95(3):243–258. doi:10.1016/0376-7388(94)00130-8

Wang S, Guillen G, Hoek EMV (2005) Direct observation of microbial adhesion to membranes. Environ Sci Technol 39(17):6461–6469

Wang Z, Wu Z, Yin X, Tian L (2008) Membrane fouling in a submerged membrane bioreactor (MBR) under sub-critical flux operation: membrane foulant and gel layer characterization. J Membr Sci 325(1):238–244

Wang, XM, Li XY, Shih K (2011) In situ embedment and growth of anhydrous and hydrated aluminum oxide particles on polyvinylidene fluoride (PVDF) membranes. J Membrane Sci 368 (1–2):134–143

Wang D, Zou W, Li L, Wei Q, Sun S, Zhao C (2011a) Preparation and characterization of functional carboxylic polyethersulfone membrane. J Membr Sci 374(1–2):93–101. doi:10. 1016/j.memsci.2011.03.021

Wang XM, Li XY, Shih K (2011b) In situ embedment and growth of anhydrous and hydrated aluminum oxide particles on polyvinylidene fluoride (PVDF) membranes. J Membr Sci 368(1–2):134–143. doi:10.1016/j.memsci.2010.11.038

Watanabe Y, Kimura K, Itonaga T (2006) Influence of dissolved organic carbon and suspension viscosity on membrane fouling in submerged membrane bioreactor. Sep Sci Technol 41 (7):1371–1382. doi:10.1080/01496390600633832

Wenten IG (1995) Mechanisms and control of fouling in crossflow microfiltration. Filtr Sep 32 (3):252–253. doi:10.1016/S0015-1882(97)84049-9

Wicaksana F, Fane AG, Chen V (2006) Fibre movement induced by bubbling using submerged hollow fibre membranes. J Membr Sci 271(1–2):186–195. doi:10.1016/j.memsci.2005.07.024

Wiesner M, Veerapaneni S, Brejchova D (1992) Improvement in microfiltration using coagulation pretreatment. In: Nice F, Klute R, Hahn H (eds) Proceedings of the fifth Gothenburg symposium on chemical water and waste water treatment, NewYork. Springer, pp 20–40

Willems P, Kemperman AJB, Lammertink RGH, Wessling M, van Sint Annaland M, Deen NG, Kuipers JAM, van der Meer WGJ (2009) Bubbles in spacers: direct observation of bubble behavior in spacer filled membrane channels. J Membr Sci 333(1–2):38–44. doi:10.1016/j. memsci.2009.01.040

Winzeler HB, Belfort G (1993) Enhanced performance for pressure-driven membrane processes: the argument for fluid instabilities. J Membr Sci 80(1):35–47. doi:10.1016/0376-7388(93) 85130-O

Wisniewski C, Grasmick A (1998) Floc size distribution in a membrane bioreactor and consequences for membrane fouling. Colloids Surf, A 138(2–3):403–411

Wu H (2012) Improving the anti-fouling and fouling release of PVDF UF membrane by chemically modified SiO_2 nanoparticles, New South Wales, Sydney, Australia

Wu J, Huang X (2010a) Use of azonation to mitigate fouling in a long-term membrane bioreactor. Bioresour Technol 101:6019–6027

Wu J, Huang X (2010b) Use of ozonation to mitigate fouling in a long-term membrane bioreactor. Bioresour Technol 101(15):6019–6027

Wu D, Howell JA, Field RW (1993) Pulsatile flow filtration of yeast cell debris: influence of preincubation on performance. Biotechnol Bioeng 41(10):998–1002

Wu Z, Wang Q, Wang Z, Ma Y, Zhou Q, Yang D (2010) Membrane fouling properties under different filtration modes in a submerged membrane bioreactor. Process Biochem 45(10):1699–1706

Xiao K, Wang X, Huang X, Waite TD, Wen X (2011) Combined effect of membrane and foulant hydrophobicity and surface charge on adsorptive fouling during microfiltration. J Membr Sci 373(1–2):140–151. doi:10.1016/j.memsci.2011.02.041

Xie RJ, Gomez MJ, Xing YJ, Klose PS (2004) Fouling assessment in a municipal water reclamation reverse osmosis system as related to concentration factor. J Environ Eng Sci 3 (1):61–72

Xu P, Bellona C, Drewes JE (2010) Fouling of nanofiltration and reverse osmosis membranes during municipal wastewater reclamation: Membrane autopsy results from pilot-scale investigations. J Membrane Sci 353(1–2):111–121

Yamamoto K, Hiasa M, Mohmood T, Matsuo T (1989) Direct solid-liquid separation using hollow fiber membrane in an activated-sludge aeration tank. Water Sci Technol 21:43–54

Yamamura H, Chae S, Kimura K, Watanabe Y (2007a) Transition in fouling mechanism in microfiltration of a surface water. Water Res 41(17):3812–3822

Yamamura H, Kimura K, Watanabe Y (2007b) Mechanism involved in the evolution of physically irreversible fouling in microfiltration and ultrafiltration membranes used for drinking water treatment. Environ Sci Technol 41(19):6789–6794

Yamamura H, Okimoto K, Kimura K, Watanabe Y (2007c) Influence of calcium on the evolution of irreversible fouling in microfiltration/ultrafiltration membranes. J Water Supply Res Technol AQUA 56(6–7):425–434

Yamanoi I, Kageyama K (2010) Evaluation of bubble flow properties between flat sheet membranes in membrane bioreactor. J Membr Sci 360(1–2):102–108. doi:10.1016/j.memsci.2010.05.006

Yang Q, Chen J, Zhang F (2006a) Membrane fouling control in a submerged membrane bioreactor with porous, flexible suspended carriers. Desalination 189 (1-3 SPEC. ISS.):292-302. doi:10.1016/j.desal.2005.07.011

Yang W, Cicek N, Ilg J (2006b) State-of-the-art of membrane bioreactors: Worldwide research and commercial applications in North America. J Membr Sci 270(1–2):201–211

Yao M, Zhang K, Cui L (2010) Characterization of protein-polysaccharide ratios on membrane fouling. Desalination 259(1–3):11–16

Yao M, Ladewig B, Zhang K (2011) Identification of the change of soluble microbial products on membrane fouling in membrane bioreactor (MBR). Desalination 278(1–3):126–131

Yeom IT, Nah YM, Ahn KH (1999) Treatment of household wastewater using an intermittently aerated membrane bioreactor. Desalination 124(1–3):193–204. doi:10.1016/S0011-9164(99)00104-6

Yeon KM, Park JS, Lee CH, Kim SM (2005) Membrane coupled high-performance compact reactor: a new MBR system for advanced wastewater treatment. Water Res 39(10):1954–1961. doi:10.1016/j.watres.2005.03.006

Yiantsios SG, Sioutopoulos D, Karabelas AJ (2005) Colloidal fouling of RO membranes: an overview of key issues and efforts to develop improved prediction techniques. Desalination 183(1–3):257–272

You HS, et al (2005) A novel application of an anaerobic membrane process in wastewater treatment, in Water Sci Technol pp 45–50

You HS, Huang CP, Pan JR, Chang SC (2006) Behavior of membrane scaling during crossflow filtration in the anaerobic MBR system. Sep Sci Technol 41(7):1265–1278

Yu HY, Hu MX, Xu ZK, Wang JL, Wang SY (2005a) Surface modification of polypropylene microporous membranes to improve their antifouling property in MBR: NH3 plasma treatment. Sep Purif Technol 45(1):8–15. doi:10.1016/j.seppur.2005.01.012

Yu HY, Xie YJ, Hu MX, Wang JL, Wang SY, Xu ZK (2005b) Surface modification of polypropylene microporous membrane to improve its antifouling property in MBR: CO_2 plasma treatment. J Membr Sci 254(1–2):219–227. doi:10.1016/j.memsci.2005.01.010

Zeng Y (2007) Membrane bioreactor technology. National Defense Industry press, Beijing, China

Zhang Y, Bu D, Liu C, Luo X, Gu P (2004) Study on retarding membrane fouling by ferric salts dosing in membrane bioreactors. In: Proceedings of the water environment-membrane technology conference, Seoul, Korea

Zhang S, Qu Y, Liu Y, Yang F, Zhang X, Furukawa K, Yamada Y (2005) Experimental study of domestic sewage treatment with a metal membrane bioreactor. Desalination 177(1–3):83–93. doi:10.1016/j.desal.2004.10.034

Zhang HM, Xiao JN, Cheng YJ, Liu LF, Zhang XW, Yang FL (2006a) Comparison between a sequencing batch membrane bioreactor and a conventional membrane bioreactor. Process Biochem 41(1):87–95. doi:10.1016/j.procbio.2005.03.072

Zhang JS, Chuan CH, Zhou JT, Fane AG (2006b) Effect of sludge retention time on membrane bio-fouling intensity in a submerged membrane bioreactor. Sep Sci Technol 41(7):1313–1329

Zhang K, Cui Z, Field RW (2009) Effect of bubble size and frequency on mass transfer in flat sheet MBR. J Membr Sci 332(1–2):30–37. doi:10.1016/j.memsci.2009.01.033

Zhang H, Gao J, Jiang T, Gao D, Zhang S, Li H, Yang F (2011a) A novel approach to evaluate the permeability of cake layer during cross-flow filtration in the flocculants added membrane bioreactor. Bioresour Technol 102:11121–11131

Zhang K, Wei P, Yao M, Field R, Cui Z (2011b) Effect of the bubbling regimes on the performance and energy cost of flat sheet MBRs. Desalination 283:221–226

Zhang K, Wei P, Yao M, Field RW, Cui Z (2011c) Effect of the bubbling regimes on the performance and energy cost of flat sheet MBRs. Desalination 283:221–226. doi:10.1016/j. desal.2011.04.023

Zhang M, Zhang K, De Gusseme B, Verstraete W (2012) Biogenic silver nanoparticles (bio-Ag 0) decrease biofouling of bio-Ag 0/PES nanocomposite membranes. Water Res 46(7):2077–2087

Zhang J, Zhang M, Zhang K (2014) Fabrication of poly(ether sulfone)/poly(zinc acrylate) ultrafiltration membrane with anti-biofouling properties. J Membr Sci 460:18–24

Zhao C, Xue J, Ran F, Sun S (2013a) Modification of polyethersulfone membranes—a review of methods. Prog Mater Sci 58(1):76–150

Zhao YH, Qian YL, Zhu, BK, et al (2008c), 'Modification of porouspoly(vinylidene fluoride) membrane using amphiphilic polymers with different structures inphase inversion process', Journal of Membrane Science, 310(1-2), 567–76

Zhou R, Ren PF, Yang HC, Xu ZK (2014) Fabrication of antifouling membrane surface by poly (sulfobetaine methacrylate)/polydopamine co-deposition. J Membr Sci 466:18–25

Zhu X, Bai R, Wee KH, Liu C, Tang SL (2010) Membrane surfaces immobilized with ionic or reduced silver and their anti-biofouling performances. J Membr Sci 363(1–2):278–286

Zsirai T, Buzatu P, Aerts P, Judd S (2012) Efficacy of relaxation, backflushing, chemical cleaning and clogging removal for an immersed hollow fibre membrane bioreactor. Water Res 46 (14):4499–4507. doi:10.1016/j.watres.2012.05.004

Zularisam AW, Ismail AF, Salim R (2006) Behaviours of natural organic matter in membrane filtration for surface water treatment—a review. Desalination 194(1–3):211–231

Chapter 4
Surface Modification of Polyethersulfone Membranes

Polyethersulfone (PES) membranes have been broadly used in scientific research and industrial processes due to their outstanding properties, such as high mechanical strength, high thermal stability, good chemical resistance and membrane forming ability. They have also been used in a cross range of applications, particularly wastewater treatment. This membrane is intrinsically hydrophobic in nature. Therefore, this property make PES membranes more prone to fouling, so surface modification techniques are an excellent pathway to improve hydrophobicity and enhance antifouling performance. In this chapter, we present different techniques to alleviate membrane fouling. This chapter sheds light specifically on surface modification of membranes. In this chapter, various types of modification techniques are outlined, including bulk modification, blending, grafting procedures and coating methods.

4.1 Introduction

Since the early 1960s and the emergence of widespread industrial membrane separations, there has been extensive research into alleviating membrane fouling, especially in membrane bioreactors (Kochkodan and Hilal 2015). An extremely wide range of approaches have been explored, including pre-treatment of feed (flocculation), optimisation of operating conditions, gas sparging, pulsatile, aggressive and frequent cleaning of the membrane, radiation and membrane modification (Baker 2012; Belfort et al. 1994; Chae et al. 2006; Hai et al. 2008; Hilal et al. 2005; Le-Clech et al. 2006a; b, c; McCloskey et al. 2012; Ridgway et al. 1984; Rosenberger et al. 2002; Sheikholaslami 1999; Sheikholeslami 1999; Wu et al. 2008; Wu and Huang 2008).

Among these methods, surface modification of membrane is considered to be an effective technology or an effective way to reduce the interactions between membrane surface and foulants, thereby reduce membrane fouling, enhancing membrane

© Springer Nature Singapore Pte Ltd. 2017
B. Ladewig and M.N.Z. Al-Shaeli, *Fundamentals of Membrane Bioreactors*,
Springer Transactions in Civil and Environmental Engineering,
DOI 10.1007/978-981-10-2014-8_4

longevity and performance, and also reducing the adsorption of protein on a membrane surface (Liu and Kim 2011; McCloskey et al. 2012; Shannon et al. 2008). As noted by (Moghimifar et al. 2014), surface modification of membrane is one of the techniques used to modify the membrane surface by increasing the hydrophilicity of the membrane and minimising membrane fouling to a minimum. In general, it is evident that the increase in membrane surface hydrophilicity enhances antifouling capability because most of the natural biopolymers (i.e. proteins) exhibit hydrophobic properties (Li et al. 2014). Theoretically, hydrophilic surfaces have the capability to display a buffer layer composed of water molecules and prevent foulants from depositing on membrane surface or at membrane pores which results in a stable and higher water flux (Peeva et al. 2012).

The surface modification of membrane may be performed by two distinct groups of polymer surface modifications: the first group is referred as chemical modification and the second group is physical modification (Mittal 2009). Physical modification of a polymer surface involves exposure of polymer to plasma, flame, radiation and ion beams method and is time-dependent (Chu et al. 2005). Chemical modification of a polymer surface necessitates or requires chemical reaction between polymer surface and functional groups and the modified polymer surface is reasonably stable over time.

Surface modification of membranes includes generation of two layers (made of two different polymeric materials) on the membranes. The first layer (thin layer) controls the flux, selectivity and adsorption of solute whilst the second layer (thick substrate) provides mechanical strength and chemical stability.

In term of protein separation or concentration using membrane processes, surface modification of polymeric membranes mostly takes the form of introducing hydrophilic functional groups or incorporating hydrophilic layers onto the surface of membrane. Examples of surface modifications aimed at improving surface resistance to adsorption of protein and membrane permeation property include the introducing of hydrophilic polymers or inorganic fillers through blending (Rajagopalan et al. 2004; Van Der Bruggen 2009a; Wang et al. 2006a, b; Zhao et al. 2011, 2013b), coating (Razmjou et al. 2011a; Brink et al. 1993; Hilal et al. 2005; Wei et al. 2005; La et al. 2011; Yu et al. 2011) and surface grafting (Kim et al. 2002; Ulbricht and Riedel 1998; Yune et al. 2011; Kang et al. 2007; Liu et al. 2008; Rahimpour 2011b; Rana and Matsuura 2010; Ulbricht and Belfort 1996; Ulbricht et al. 1996; Revanur et al. 2007a), self-assembly monomers (Luo et al. 2005). These techniques have successfully enhanced membrane surface properties, such as improving membrane permeation, improving mechanical strength, hydrophilicity and fouling resistance (Zhu et al. 2014). However, they also have some shortcomings or limitations hindering their further popularity in membrane industry, for instance, coating method can be simply applied at industrial scale but the coating layer is instable (easily removable in use. In contrast, grafting is easily adhering but not easy to apply at industrial use. They usually need substantial post treatment process, leading to high cost of membrane fabrication costs (Zhu et al. 2014).

To conclude, surface modification of membrane is an effective strategy to modify the membrane surface and enhance its surface characteristics such as increase membrane hydrophilicity, increase water flux and minimise adhering inorganic, organic and biological materials on the membrane surface. Bear in mind, inorganic, organic and biological materials are the main cause of occurrence fouling in membrane bioreactors.

4.2 Surface Modification

As aforementioned, PES membrane is intrinsically hydrophobic in nature. Numerous research studies for example (Khulbe et al. 2010a, b; Van Der Bruggen 2009a) stated that membrane fouling is related directly to hydrophobic membranes. Therefore, this membrane needs to be modified in order to alleviate membrane fouling to a certain extent. Three approaches have been used to modify PES membrane: (1) bulk modification of the prepared PES membrane, and then to prepare modified membrane; (2) surface modification of prepared PES membrane; and (3) blending (it is regarded as a surface modification). Membrane modification method can be used to find a compromise between hydrophobicity and hydrophilicity, by localising hydrophilic functional groups specifically in the membrane pores. They have a great impact on membrane permeation, morphology of membrane, and provide the membrane with stimuli-responsivity and better blood compatibility (Zhao et al. 2013a). The first membrane modification process which was reported in the literature is annealing of porous membranes through heat treatment (Khayet et al. 2003; Pinnau and Freeman 2000). The porous membrane was used in different applications such as reverse osmosis, gas separation and ultrafiltration membranes.

Table 4.1 shows a number of membrane modification methods such as annealing with heat treatment, solvent treatment, surface coating, chemical treatment, and their application in an industrial area.

Van Der Bruggen (2009a) studied the chemical and physicochemical modification methods, which were used to enhance membrane hydrophilicity of PSF/PES nanofiltration (NF) membranes. In case of increasing hydrophilicity of membranes, modification of membranes can be implemented in different techniques. Physical or chemical modification methods after the membrane is prepared or formed can endow the membranes with hydrophilic surfaces. Such modification processes include (1) graft polymerization process (hydrophilic monomers are chemically attached to the surface of membrane) (2) plasma treatment process (incorporate the functional graft chains on the surface of membrane); and (3) physical pre-adsorption of hydrophilic groups to the surface of membrane.

Obviously, various modification methods have been used to modify PES membranes including physical processes such as blending and surface-coating processes (Li et al. 2010; Liu et al. 2009; Ran et al. 2011; Reddy et al. 2003; Wang et al. 2009b), and chemical processes including photo-induced grafting (Zhao et al. 2003), gamma ray (Filho and Comes 2006; Deng et al. 2009) and electron beam-induced grafting (Keszler et al. 1991; Schulze et al. 2010), plasma treatment and plasma-induced grafting (Batsch et al. 2005; Gancarz et al. 1999b; Van Der Bruggen 2009a), thermal-induced grafting and immobilisation (Fang et al. 2009,

Table 4.1 Common membrane modifications

Modification method	Function	Application
Annealing – Heat treatment – Solvent treatment	– Elimination of membrane defects – Control of pore size	RO, GS, UF
Solvent exchange	– Elimination of membrane defects	GS, UF
Surface coating	– Elimination of membrane defects – Melioration of fouling resistance	GS, RO, NF,UF
Chemical treatment – Fluorination – Cross-linking – Pyrolysis	Improvement of flux and selectivity Improvement of chemical resistance Improvement of flux and selectivity	GS UF RO, GS, PV

GS gas separation, *RO* reverse osmosis, *UF* ultrafiltration, *NF* nanofiltration, *PV* pervaporation
(Pinnau and Freeman 2000)

2010; Kroll et al. 2007; Liu et al. 2009; Shi et al. 2010a, b), surface-initiated atom
transfer radical polymerization (Li et al. 2009c, d, e, f), and so on (Cao et al. 2010;
Liu et al. 2008; Rana and Matsuura 2010; Tari et al. 2010; Sun et al. 2003). These
methods have successfully modified the surface of polymeric membranes, such as
PES, PVDF, and PP and so on. The function of these methods is to make the
membrane super hydrophilic and reduce membrane fouling to a great extent.

With the development in membrane modification process, reversible
addition-fragmentation chain transfer polymerization (RAFT) and click chemistry
methods are new techniques used recently to modify PES membranes. They are
discovered by CSIRO.

4.3 Bulk Modification

Bulk modification methods are usually very simple methods because they are used
to a polymer solution, and not on a membrane surface (Peyravi et al. 2012).
However, the entire or whole membrane is modified in this way, therefore leading
to a lower net effect and, possibly, an enhanced effect of swelling in the structure of
the resulting membrane (Zhu et al. 2007a, b). Sulfonation and carboxylation are the
most commonly reported methods in many journal articles. But sulphonation is one
of the membrane modification methods, which used widely in industrial section
(Wang et al. 2009a; Rahimpour et al. 2010a; Cao et al. 2010). Sulfonation of PES
membrane is an electrophilic reaction, in which the negatively charged sulfonic acid
groups are introduced onto PES membranes, resulting in replacing the hydrogen
atom with sulfonic acid groups as shown in Fig. 4.1 (Van Der Bruggen 2009a). As
can be seen from the figure, the sulfonic group is localised in ortho positions on the
aromatic rings of PES polymer because this electron donating oxygen atom which
activates ortho position (Blanco et al. 2001; Iojoiu et al. 2005). Sulphonation
method is affected by the electron donating properties of a polymer and elec-
trophilicity of the sulphonating agent. However, PES matrix is thoroughly hard to

sulphonate due to the electron withdrawing effect of the sulfonic linkages, which deactivates the adjacent aromatic rings for electrophilic substitution. PES membranes generated by sulphonation reaction are super hydrophilic, have excellent antifouling property, proton conductivity and better selectivity. The sulphonation of PES has been described in many journal articles with different sulphonating agents and solvents used: SO_3/CH_2Cl_2 [5, 7], $ClSO_3H/CH2Cl2$ [8], SO_3-triethylphosphate (TEP)/CH_2Cl_2 [9], oleum/concentrated sulfuric acid [10].

In contrast, carboxylation is a practical method, which involves addition of carboxylic groups to polymeric PES membrane matrix, in which carboxyl group substitutes the hydrogen atom at an aromatic hydrocarbon. In carboxylation method, PES can be oxidised or acetylated at specific conditions as explained in Fig. 4.2. PES membranes generated by carboxylation, have a great influence on the hydrophilicity of the polymer, which was verified by measurements of water adsorption and measurements of contact angle (Zhao et al. 2013b, a). Introducing carboxylic groups to PES matrix, as shown in Fig. 4.2, could increase the hydrophilicity of membranes.

Apart from sulphonation and carboxylation methods, nitration is another method used to introduce amine groups to the polymeric PES membrane to modify its surface. Van Der Bruggen (2009a) states that the limitation of bulk modifications can be solved through blending method, which is the same to the use of copolymers but using a physical route. This also allows the use of "simple" polymers. Recently, different functional groups were introduced to PES membranes. PES was firstly sulfonated by chlorosulfonic acid ($ClSO_3$) at 0 °C for 2 h to form sulfonated PES. The degree of sulphonation was about 15 %. Then sulfonated PES was chlorinated by phosphorous pentachloride to form $SPES/SO_2Cl$ as presented schematically in Fig. 4.3 (Bai et al. 2010). Sulphonated PES is more reactive toward amine group, hydroxyl group, carboxyl group and methyl group as shown in Fig. 4.3.

Shi et al. (2010) grafted methacylic acid (MAA) onto polymeric PES membrane network in heterogeneous reaction, using benzoyl peroxide (BPO) as a chemical initiator. The polymerisation process was implemented in an aqueous medium. Because of the conformational change of PMAA chains with environmental pH, membrane cast from PMAA-g-PES exhibited reversible pH dependent permeability as the pH value of feed solution was varied (Zhao et al. 2013a).

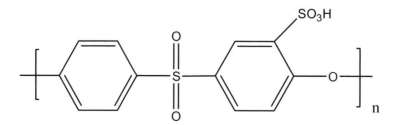

Fig. 4.1 Polyethersulfone, with the sulfonic acid group on the benzene ring

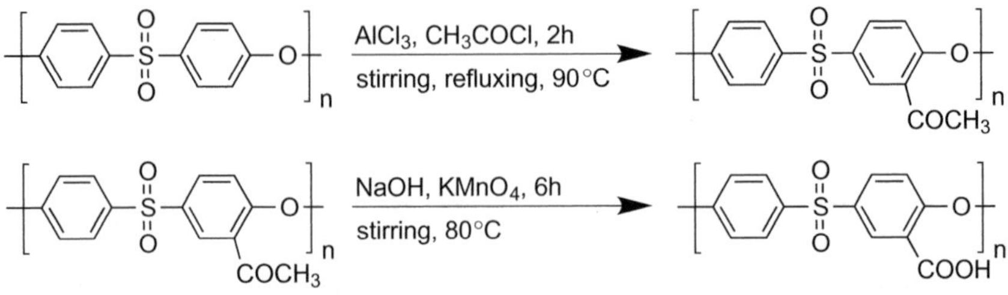

Fig. 4.2 Fabrication of polyethersulfone membranes by carboxylation method (adapted from Wang et al. 2011)

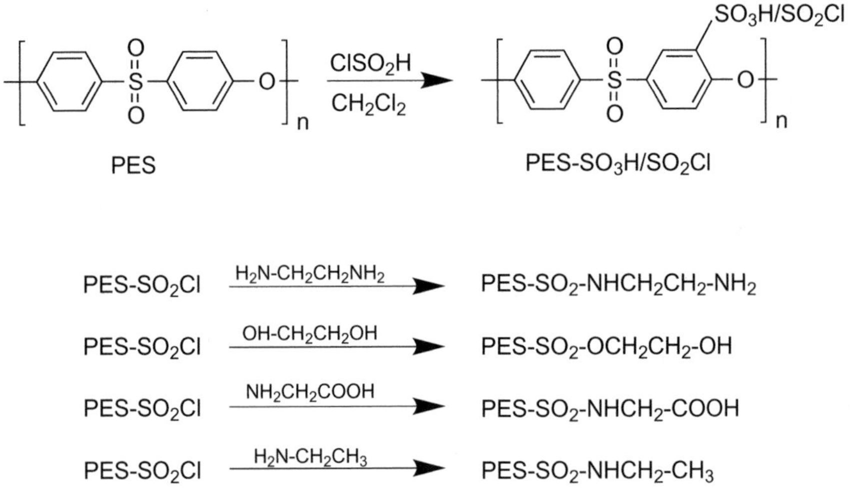

Fig. 4.3 Synthesis of PES with different functional groups (Adapted from Zhao et al. 2013a; Bai et al. 2010)

4.4 Physical Modification of PES

Physical modification of PES usually involves one of two distinctive approaches, either the blending technique or the surface coating technique.

4.4.1 Blending

Blending is the simplest method and most effective modification process to modify polymeric membranes (both flat-sheet and hollow fiber membranes PES membranes), since the results might not very effective at certain time (Zhao et al. 2013a). Moreover, it is regarded as the most practical way, which can be used to an industrial scale production. PES membranes are very easy to modify in this method. This method is often employed to obtain the desired functional properties along with the membrane preparation. Thus, both inside pores of membranes and the membrane surfaces can be modified simultaneous through the interactions between the basic polymer and additives. This can enable the prepared membrane to have the comprehensive properties of the blend materials (Venault et al. 2014; Bottino et al. 2000; Ying et al. 2002; Madaeni et al. 2011).

By directly blending hydrophilic polymers additives, for example polyvinyl-pyrrolidone (PVP) (Barzin et al. 2004; Dang et al. 2008) and polyethyleneglycol (PEG) (Wang et al. 2006a, b, c; Su et al. 2008; Dang et al. 2008) into membrane bulk, the membrane hydrophilicity increased, the antifouling performance and blood compatibility are also increased. PVP and PEG are homopolymeric polymers that used as pore—forming agents. These additives are water-soluble and can be leached out from hydrophobic PES membranes during membrane preparation and operation processes, leading to deterioration of membrane performance. Therefore, cross-linking with additives or covalent grafting has been proposed to tackle the issue of leaching out these additives (Park et al. 2006; Tripathi et al. 2012). Tripathi et al. (2014) prepared highly hydrophilic and fouling resistant porous membranes by covalent cross-linking of sulfonated PES with amino terminated poly ethylene glycol. PES was sulfonated and cross-linked with poly (ethyleneglycol) bis-(3-aminopropyl)terminated via sulfonamide linkage using1,1'-carbonyldiimidazole (CDI). The cross-linked membranes showed superior water permeation flux, rejection, and antifouling performance in comparison to the neat membranes. The protein adsorption on membrane surface was about10-foldless than that of the control PES membrane The results reveal that, hydrophilicity was increased due to the existence of PEG, leading to the 3-fold increase in water permeation flux and more than 90 % protein rejection efficiency.

Another approach to avoid the leaching out of these additives is using Amphiphilic copolymers. Recently, these polymers are synthesised and used as additives to fabricate PES membranes (Zhu et al. 2008a, b).

Various blending methods have been devoted or performed to modify PES membranes to enhance their antifouling properties and improve the hydrophilicity. These methods include blending with hydrophilic nanoparticles, blending with homopolymeric membranes (e.g. PVP and PEG), blending with amphiphilic polymers, and blending with zwitterionic polymers, and blending with surface modifying macromolecules.

Blending with hydrophilic nanoparticles has become a new domain of interest in membrane separation technology due to its ease and simplicity (Jamshidi Gohari et al. 2014). The resulting membranes have unique properties, including excellent separation performance, reasonable operation under harsh conditions (Yang et al. 2006), adaptability to milder environments, good thermal and chemical stability, and excellent antifouling properties (Molinari et al. 2002; Akar et al. 2013). The purpose of incorporating nanoparticles into polymeric membranes is to improve hydrophilicity and to improve pore formation and interconnectivity (Rahimpour and Madaeni 2010; Rahimpour et al. 2010b, c). Nanoparticles have indeed unique properties such as chemical properties; thermal and mechanical stable properties, electrical properties, small size, large surface area and strong additives. Therefore, they have attracted a great deal of attention recently. Akar et al. (2013) states that incorporating nanoparticles with membrane results in improved membrane surface properties (regarding to chemical and thermal resistance), improved separation performance, mechanical properties and antifouling properties. Therefore, various nanoparticles have been utilised in fabricating PES membrane, including, titanium dioxide (TiO_2) (Arthanareeswaran et al. 2008; Luo et al. 2005, 2011; Razmjou et al. 2011a, b; Devrim et al. 2009; Sotto et al. 2011), mesoporous silica (MS) (Huang et al. 2012; Yu et al. 2009b; Yin et al. 2012), modified multi-walled carbon nanotubes (Daraei et al. 2013b), silver (Huang et al. 2012), magnesium hydroxide Mg $(OH)_2$ (Dong et al. 2012), calcium carbonate nanoparticles (CCNP) (Nair et al. 2013) and Carbon nanotube (CNT) (Majeed et al. 2012; Ajmani et al. 2012) and HNTs (Liu et al. 2009).

Chen et al. (2013) investigated the effects of halloysite nanotubes—chitosan (CHI)—Ag nanoparticles on the antifouling properties and performance of PES membranes and the results demonstrated that the adding halloysite nanotubes—CHI—Ag nanoparticles into PES membranes could lead to an increase in the hydrophilicity of the modified membrane and then resulted in excellent both membrane permeation and flux recovery ratio (FRR) in comparison to the neat PES membrane.

Razmjou et al. (2012) investigated the effect of TiO_2 nanoparticles on the morphology, performance and antifouling properties of PES membrane. They concluded that incorporating TiO_2 nanoparticles into PES membranes, indicating good membrane hydrophilicity, lower contact angle and higher FRR.

Dong et al. (2012) point out that Mg $(OH)_2$ nanoparticles are very efficient in preventing flux declines, causing the hybrid membrane to show excellent antifouling performance, and higher hydrophilicity than the unmodified membrane. The performance of membranes is enhanced due to the existence of—OH groups on the surface of membrane.

Nair et al. (2013), also investigated the effect of OH groups on the hydrophilicity of membranes when the hydroxyl group was formed from CCNP to Ca–OH in an aqueous medium.

Jamshidi Gohari et al. (2014) investigated the introducing of hydrous manganese dioxide (HMO) nanoparticles to PES ultrafiltration membrane. They concluded that these nanoparticles have higher capability for enhancing the properties of UF membranes, especially their antifouling properties. HMO nanoparticles offer several advantages, such as microporous structure, large surface area and can be readily prepared through redox reaction (Taffarel and Rubio 2010; Teng et al. 2009).

Razali et al. (2013) investigated the effect of incorporating polyaniline nanoparticles as polymeric additives to PES membranes. They stated that these nanoparticles have the capability to enhance the hydrophilic properties and permeability of the substrate membrane. Polyaniline nanoparticles are used to achieve super hydrophilic surfaces due to their high-surface energy and hydrophilic properties (Fan et al. 2008).

Recently, Zinadini et al. (2014) investigated the blending of graphene oxide nanoparticles (GO) onto PES based-UF membrane. They demonstrated that these inorganic nanoparticles have had the best antifouling property when 0.5 wt% concentrations added to the dope solution.

Huang et al. (2012) used MS particles as an advanced antifouling membrane material. The membrane hydrophilicity and antifouling performance were improved significantly. Li et al. (2014) employed Polydopamine (PD) coating and PD-graft-poly(ethylene glycol) (PD-g-PEG). The results showed that the modified membrane had a fine mechanical stability, and the PD-g-PEG modified membrane had a better chemical stability. The membrane hydrophilicity was increased according to the measurements of contact angle. In comparison to the pristine membranes, the modified membranes adsorbed less BSA under the same condition. Furthermore, PD-g-PEG modified membrane seems to have less adsorptive fouling potential than PD-coating membrane. Generally, PD-g-PEG modification should be more promising in industrial application or scale.

Modified PES UF membranes by SiO_2@N-Halamine. The membrane hydrophilicity was significantly enhanced after introducing SiO_2@N-Halamine. The filtration results indicated that the permeation properties of the hybrid membranes were significantly superior in compared to the control PES membrane. The water flux of the hybrid membranes increased with the additional amount of SiO_2@N-Halamine increased. When the amount of SiO_2@N-Halamine was 5 %, the membrane permeation of hybrid membrane reached the maximum at 384.4 L $m^{-2} h^{-1}$. Furthermore, the hybrid membranes showed good antifouling and antibacterial properties, which might extend the usage of PES in the application of water treatment and could make some potential contributions to membrane antifouling.

Pang et al. (2014) developed a simple in situ method for the preparation of hydrous zirconia ZrO_2/PES UF. They merged an ion-exchange method with a classic immersion precipitation method. The solution of hydrous zirconia (ZrO_2) was prepared by addition of anion-exchange resin in N, N-dimethylformamide (DMF) solvent containing zirconyl chloride. The results indicated that the ZrO_2 nanoparticles were dispersed well into PES matrix and the diameter of the prepared nanoparticles was about 5–10 nm. The hydrophilicity of the hybrid membranes was improved significantly. The adsorption of protein was reduced as a result of incorporating ZrO_2 nanoparticles in the PES matrix.

Zhao et al. (2015) developed a new method for modifying PES/ZnO UF membranes. Nano-ZnO was coated with PVP polymer using DMF as solvent. The obtained ZnO-DMF dispersion was blended as an additive to fabricate PES/ZnO membranes via wet phase separation. The results showed that the antifouling performance was improved. The thermal stability of the hybrid membranes was also enhanced in comparison to the pristine PES membranes.

Martín et al. (2015) modified PES UF membranes by incorporating mesostructured modified silica particles in PES membranes. The antifouling properties of the treated membranes were improved, particularly against irreversible fouling. Multi-run fouling experiments of the hybrid membranes verified the stability of membrane permeation after introducing mesostructured modified silica.

Xiang et al. (2014) prepared ionic-strength-sensitive membrane through in situ cross-linked polymerization of sulfobetaine methacrylate (SBMA) in PES solution and a liquid–liquid phase separation process. The membrane with high amounts of PSBMA showed an obvious ionic-strength-sensitive property and ionic-strength reversibility, which are expressed by the fluxes of salt solutions. At the same time, the antifouling property and blood compatibility are increased significantly when the amount of PSBMA increased in the PES membrane.

Xie et al. (2015) prepared a novel zwitterionic glycosyl modified PES UF membranes through coupling in situ cross-linking polymerization process with phase inversion process. The results showed the modified membranes have an excellent antifouling performance, and the FRR could reach almost 100 %. Menawhile, the blood compatibility of the membranes was determined by protein adsorption experiment; platelet adhesion activated partial thromboplastin time (APTT), and thrombin time (TT). The results indicated that the surface modified PES membranes had good antifouling performance and blood compatibility (anticoagulant).

Duan et al. (2015) modified PES ultrafiltration membrane by blending with N-halamine grafted HNTs (halamine@HNTs), which is anticipated to improve the antibacterial property and the permeability of the membranes. The results indicated that the membrane hydrophilicity was enhanced significantly after adding N-halamine@HNTs. The membrane permeability (Water flux) of the hybrid membranes could reach as high as $248.3 \, \mathrm{L \, m^{-2} \, h^{-1}}$ when the amount of N-halamine@HNTs was 1.0 wt%. Additionally, the antibacterial experiment indicated that the hybrid membranes showed good antibacterial activity against $E. \, coli$ (Table 4.2).

Ultimately, one of the most significant issues with blending nanoparticles into polymer matrix is the agglomeration phenomenon. An agglomeration phenomenon causes the non-uniform distribution of particles in the membrane and the instability of the casting solution. This will lead to a big change in microstructure, topography and performance and also the potential decreasing of antifouling ability of nanoparticles. There have been a number of articles reporting that agglomeration phenomena is due to the nature properties of nanoparticles (e.g. small size and high surface energy) and the bad compatibility with hydrophobic PVDF bulk (Kang and Cao 2014). Agglomeration phenomena can be reduced by three approaches:

1. Functionalisation of nanoparticles or surface modification of nanoparticles (Zhang et al. 2014; Razmjou et al. 2011b; Shi et al. 2013)
2. In situ formation of nanoparticles via sol-gel method (Yu et al. 2009b)
3. Creating the bridge to improve the interaction between nanoparticles by the addition of a third component. Therefore, the nanoparticles have to be carefully selected to ensure their dispersion in the membrane matrix to avoid agglomeration issues.

Table 4.2 Illustrates some of the recent research studies that have been conducted PES membranes through blending technique using different nanoparticles (Jamshidi Gohari et al. 2014)

Polymer	Nanoparticles	Pure water flux (L/m²h)	Protein flux (L/m²h)	Rejection (%)	Contact angle (°)	Flux recovery (%)	Reference
PES							
PES	Halloysite nanotubes-chitosan-Ag NIPS	357.6 at 1 bar	0.75 normalised flux	N/A	55	97.6	Chen et al. (2013)
PES	Functionalised multi-walled carbon nanotube	25 at 4 bar	5–10 (whey protein)	97 (whey protein)	49.8	95	Daraei et al. (2013a)
PES	Mesoporous silica	180.2 at 2 bar	75.8	96.1 BSA	56.6	76.2	Yin et al. (2012)
PES	Ag-loaded sodium zirconium phosphate NPs	100.6 at 1 bar	51	96.7 BSA	58.4	78.4	Huang et al. (2012)
PES	PANi/PMA NPs	320 at 2 bar	225	99 BSA	72.3	72.8	Taffarel and Rubio (2010)
PES	Copper particles	69 (0.5–1.5) bar	N/A	86.3 BSA	55.3	76.3	Akar et al. (2013)
PES	Biogenic silver NPs	520 at 2 bar	N/A	N/A	51.4	N/A	Teng et al. (2009)
PES	TiO₂	55 at 1 bar	N/A	95 (dextran 500 Da)	50	61.54	Razmjou et al. (2012)
PES	PES/PVP semi-interpenetrating network NPs	243.2 at 0.5 bar	50–60	93 BSA	53	94	Su et al. (2010)
PES	TiO₂	450 at 1 bar	N/A	94 (dextran 500 Da)	62	84	Mishra and Vijaya (2007)
SPES	TiO₂	113.5 at 2 bar	N/A	34.5 (PEG-400)	24.5	N/A	Luo et al. (2011)
PES	Silica	168.4 at 1.1 bar	110	96.3 BSA	52.6	70.4	Parida et al. (1981)
PES	TiO₂	102.9 at 2 bar	N/A	34.5 (PEG-5000)	19.2	N/A	Luo et al. (2005)

Surface Coatings

Surface coating is an alternative technique that has been extensively used in different applications and it may apply as liquid, gas or solid. It has been specifically used to alleviate membrane fouling in membrane bioreactors (Razmjou et al. 2012). In this technique, a thin selective layer is directly deposited as a coating layer (thin film layer) on the top surface of membrane (Zhao et al. 2013a). Surface coating method is simple and economical than other processes; however, the main challenge of coating method is instability of the coating layer and limited diffusion of the modification agent. There have been numerous studies reporting that surface coating can be simply applied successfully in industrial scale but the coating layer is readily removable in use after long usage of membranes (Li et al. 2014; Rana and Matsuura 2010). Table 4.3 summarises surface modification of polymeric membranes via surface coating techniques.

A variety of coating techniques have been described in many research articles, including chemical vapour deposition (CVD) (Ha et al. 1996; Yan et al. 1994), physical vapour deposition (PVD) (Yun and Ted Oyama 2011), chemical (solution adsorption and sol-gel method) (Luo et al. 2005; Mansourpanah et al. 2009; Rahimpour et al. 2008) and plasma spraying (Lin et al. 2012) coatings. These techniques are mostly employed for the modification of membranes, in particular ceramic membranes with gas separation applications. Each technique has limitations, which impair its widespread application for fabrication of polymeric membrane. For example CVD, PVD and plasma spraying are very expensive due to the need to work at high temperature and pressure, the need for a high vacuum, and as such they require high energy usage and the substrate geometry is limited to a flat surface (Yun and Ted Oyama 2011). However the chemical paths are much more applicable to polymer membranes as they can feasibly operate at low temperature and pressure, and consume much less energy than the previously mentioned routes.

Chemical modification of PES

Chemical modification of PES membranes involves different techniques, which can be classified as follows.

Photo-induced grafting (Photochemical-initiated graft polymerization)

Photochemical surface functionalization is an attractive approach used to modify polymeric membranes (e.g. PES). It has been widely investigated by many scholars, especially the UV grafting method (Hilal et al. 2003, 2004; Kaeselev et al. 2001; Kilduff et al. 2000; Pieracci et al. 1999, 2000; Qiu et al. 2007; Taniguchi and Belfort 2004). UV method is more applicable to modify flat-sheet membrane; it is hard to modify hollow fiber membranes, particularly to modify the internal surface of hollow fiber membrane.

Photochemical-initiated graft polymerization has several advantages and limitations. The advantages of this method are inexpensive method (low cost), operate under harsh or milder reaction conditions and low temperature may be used to the reaction; the possibility of choosing different reactive groups or monomers in the

Table 4.3 Surface modification of polymeric membranes through surface coating techniques (Rana and Matsuura 2010)

Base material	Treatment or remedy	Function of the membrane	Reference
Cross-linked polystyrene embedded into PE	Fluorinated long-chain pyridinium bromide	Sodium humate filtration; the modified membranes were prepared by deposition of fluorinated amphiphilic compound in an oriented layer of the Langmuir–Blodgett type	Speaker (1986)
PS	PEI	UF membranes, flux reduction in ovalbumin solution; hydrophilicity of the membrane appeared to be a more important factor in flux reduction than the charge	Nyström (1989)
PES	Polyurea/PU	UF membranes, flux reduction in PEG, BSA, dextran and surfactant solutions; flux enhancement are possible with the modifications	Hvid et al. (1990)
Reinforced PVC	Polypyrrole	Electrodialysis (ED) membranes, flux reduction in organic foulants; modified membranes showed excellent anti-organic fouling properties in electrodialysis	Sata (1993)
Sulphonated PS and PES	Quaternized poly(vinyl imidazole)	UF membranes, flux reduction in BSA solution, and lysozyme solutions; significant improvement in protein adsorption was observed for modified membranes at low ionic strength	Millesime et al. (1994b)
PES	Bentonite, diatomite, iron oxide, kaolinite, titanium dioxide, zeolite, etc.	UF membranes, treatment of surface water from Twenty canal, lake, and reservoir (Delft, Netherlands); pre-coating results initially in higher rate of fouling, which stabilises after several filtration cycles	Galjaard et al. (2001)
PES	Hydrophilic triblock copolymer, PEO-b-PPO-b-PEO, surfactant	Pulp and paper effluent filtration; increasing the hydrophilic characteristics of the membrane before filtration could reduce the amount of organic foulants adsorbed to the membrane	Maartens et al. (2002)

(continued)

Table 4.3 (continued)

Base material	Treatment or remedy	Function of the membrane	Reference
CA and PVDF	Phospholipid	Microfiltration (MF) membranes, flux reduction in BSA solution; phospholipid coating improved flux more in the PVDF membrane than in CA membrane	Akhtar et al. (1995)
PVDF	Polyether-*b*-PA block copolymer	UF, oil-water emulsion from metal industry; composite membrane was found to perform similar to Amicom YM30 cellulose membrane, but with lower susceptibility to fouling in the UF of oil-water waste	Nunes et al. (1995)
Zirconium oxide inorganic membrane	Quaternized poly(vinyl imidazole)	UF, flux reduction in BSA solution, and lysozyme solutions; modified presumed to be negatively charged	Millesime et al. (1994a)
CTA, PES, and PVDF	Phospholipid (MPC)	MF, flux reduction in BSA solution, yeast fermentation broth, beer, and orange juice; the coating increased the initial flux and decreased the fouling rate; the fouling was caused by mostly polysaccharide rather than protein	Reuben et al. (1995)
PES	PVA	UF membranes, flux reduction in BSA solution; improved the antifouling property of PES membrane. The flux recovery ratio increased significantly	Ma et al. (2007)
PES	PEGDA and trimethylolpropane trimethylacrylate via a thermal-induced surface cross-linking process	BSA filtration; the modified membranes were less susceptible to fouling and had grater flux recoveries after cleaning as compared to pristine membrane	Mu and Zhao (2009)
PVDF	Dip and surface flow coated chitosan	BSA adsorption and filtration; the modified membranes were observed higher flux recovery than neat membranes; membrane modified by combined dip coating and surface flow methods displayed the best antifouling properties	Boributh et al. (2009)

(continued)

Table 4.3 (continued)

Base material	Treatment or remedy	Function of the membrane	Reference
PVDF	Coated with PVA and cross-linked by glutaraldehyde vapour	BSA filtration and natural water of Grand River (Kitchener, Canada); the reduction in BSA flux solution was remarkably lowered, and higher flux and slower rate of fouling were noticed during natural water filtration by modified membrane	Du et al. (2009)
PVC	MPC copolymer (PMB)	Adhesion in platelet-rich plasma (PRP) by fluorescence micrograph; the modified membrane exhibited significant reduction in biofouling phenomena	Berrocal et al. (2002)
PS	Pluronic triblock copolymer, PEO-*b*-PPO-*b*-PEO, surfactant	Proteins adsorption, and adhesion in PRP; it was suggested that the bioinert property of PEO segments in the pluronic suppresses the adsorption of plasma proteins and platelets to the coated membranes	Higuchi et al. (2003)
PVDF	PEO-*b*-PA 12 block copolymer	UF spiral-wound membranes, motor oil-water emulsion filtration; the coated membranes exhibited significantly low-fouling properties	Freeman and Pinnau (2004)
PVDF	PVDF-*g*-POEM graft copolymer	Oleic acid-triethanol amine-water filtration; modified membranes exhibit nanoscale size selectivity with good wetting properties	Akthakul et al. (2004)
PVDF	Hydrophilic polymer	UF membranes, BSA and enzyme filtration, and adhesion in human PRP; as the antifouling properties were excellent, the membrane could be cleaned without using any cleaning agent	Wei et al. (2006b)
PVDF	Poly(cyclooctene)-*g*-PEG graft copolymer	UF, soybean oil-water emulsion filtration; the stability and lifetime of the coated membrane against oil droplet were good, and after a long run the coated membrane flux crosses over the control	Revanur et al. (2007b)

(continued)

Table 4.3 (continued)

Base material	Treatment or remedy	Function of the membrane	Reference
Microporous PP	Quarterisation cross-linking hydrophilic and positively charged coating by poly (ethylene imine), cross-linked with *p*-xylylene dichloride and quaternized by iodomethane after plasma pre-treatment	BSA and lysozyme filtration; the modified membranes resisted effectively protein fouling below the isoelectric point; furthermore, MF characteristic has been unchanged	Yang et al. (2009a, b)
PS and PVDF	Coated with poly (cyclooctane-*g*-PEG)	Oil-water emulsion; copolymers-coated films reduced fouling for water purification	Emrick et al. (2008)
Polypropylene	Coated with poly(sulfobetaine methacrylate) and polydopamine co- deposition	UF: BSA, Hgb and LYs filtration, the coated membranes exhibited significantly excellent antifouling property and water flux	Zhou et al. (2014)
PES	Coated with polydopamine (PD) and PD-graft-poly (ethylene glycol)	UF: BSA filtration; the mechanical stability of the membranes were fine and the PD-g-PEG modified membrane had a better chemical stability, the modified membrane had less flux reduction and lower adsorptive amount of BSA	Li et al. (2014)
PES	Corona air plasma and coated with TiO_2 nanoparticles	UF: significant enhancement of the surface hydrophilicity, improvement of the antifouling properties and permeation fluxes for all modified membranes, the modified membranes had a lower fouling tendency and long term flux stability in comparison to the virgin PES membranes	Moghimifar et al. (2014)
PES	Coated with Nanoporous Parylene	UF: improved biocompatibility of the PES membranes	Prihandana et al. (2012)
PES	Coated with sulfonated poly (propyleneoxide) PPO	UF: the modified membranes has lower water flux, enhancement in hydrophilicity surface, and lower membrane fouling	Singh et al. (1997)

(continued)

Table 4.3 (continued)

Base material	Treatment or remedy	Function of the membrane	Reference
PES	Preadsorption of poly (sodium 4-styrenesulfonate)	UF: the surface modified membranes shower excellent antifouling properties	Reddy et al. (2003)
PES	Coated with polydopamine-grafted polyethylene glycol	UF: The modified membranes had less flux reduction in filtration and lower adsorptive amount of BSA in isothermal adsorption tests. The PD-g-PEG modification improves the stability of the PES membrane and the adsorbability for BSA more significantly	Li et al. (2014)

grafting procedure and respective excitation wavelength; possibility of easy introduction into the final stages of a membrane manufacturing process, and high selectivity to absorb UV light without affecting the bulk polymer (Hilal et al. 2003). The disadvantages of this method is that it usually adheres at small scale but not so easily for industrial scale (Li et al. 2014). Photo-induced grafting can be achieved via two routes: either **with or without photoinitiator**. The mechanisms of photo-induced grafting without photoinitiator, on the one hand, involves the direct generation of free radicals on the backbone of the membrane polymer under UV irradiation, which react with the monomer free radical to form the grafted copolymer (Seman et al. 2012). Due to the sensitivity of PES, PS polymer toward UV irridation, the photoinitiator is not required. Different hydrophilic monomers have been grafted onto the membrane surface of PS and PES to decrease their fouling by proteins and increase their hydrophilicity, including N-Vinyl-2-pyrrolidinone (NVP), 2-hydroxyethyl methacrylate (HEMA), acrylic acid (AA), acrylamide (AAm), and 2-acrylamidoglycolic acid (AAG) (Kilduff et al. 2000; Mohd Yusof and Ulbricht 2008; Pieracci et al. 2000; Taniguchi and Belfort 2004; Yamagishi et al. 1995a, b). As stated by numerous research studies, HEMA was the most effective hydrophilic monomer to decrease both static adsorption and the membrane fouling during BSA filtration. Other hydrophilic monomers have been grafted onto PES ultrafiltration membranes via photo-irradiation process to enhance the anti-fouling property of membranes. These monomers include AA (Abu Seman et al. 2010; Gupta et al. 1992, 2009; Kouwonou et al. 2008; Malaisamy et al. 2010; Taniguchi and Belfort 2004; Taniguchi et al. 2003), 2-acrylamidoglycolic acid monohydrate (AAG) (Kaeselev et al. 2001; Taniguchi and Belfort 2004; Taniguchi et al. 2003; Wei et al. 2006b), 2-acrylamido-2-methyl-1-propanesulfonic acid (AAP) (Kaeselev et al. 2001), 2-(acryloyloxy)ethyl trimethyl ammonium chloride (AETMA) (Madaeni et al. 2011), AAm (Gupta et al. 2009), 2-acrylamido-2-methyl-1-propanesulfonic acid (AMPS) (Hilal et al. 2003; Taniguchi and Belfort

2004; Taniguchi et al. 2003), quaternary-2-dimethylaminoethyl methacrylate (qDMAEMA) (Hilal et al. 2003), HEMA (Taniguchi and Belfort 2004; Taniguchi et al. 2003; Wei et al. 2006b), methacrylic acid (MA) (Wei et al. 2006b), PEG monomethacrylate (PEGMA) (Rahimpour 2011a; Saha et al. 2009; Susanto et al. 2007; Susanto and Ulbricht 2008), N,N-dimethyl-N-(2-methacryloyloxyethyl-N-(3-sulfopropyl) ammonium betaine (Rahimpour 2011a; Susanto and Ulbricht 2007), 3-sulfopropyl methacrylate (SPMA) (Taniguchi and Belfort 2004; Taniguchi et al. 2003) and N-vinylfonnamide (NVF) (Wei et al. 2006b).

In contrast, surface modification with an added photoinitiator; the initiating radical sites are formed as a result of photoinitiator itself. These sites should be generated on the surface of membrane through the reaction of photoinitiator with the hydrogen atom in the base polymer under UV irradiation, yielding the radical sites required for grafting (Bhattacharya and Misra 2004). A schematic diagram of the photoinitiated graft polymerization method is shown in Fig. 4.4.

It should be pointed out that a number of factors affect the grafting degree of the membrane; including the method of UV grafting, the dip or immersion method (Kilduff et al. 2000; Pieracci et al. 2000) the use of a photoinitiator (Yamagishi et al. 1995b; Lee et al. 2004; Qiu et al. 2005); the application of a degassing agent, for example, nitrogen or argon (Chu et al. 2006; Liang et al. 2000; Pieracci et al. 2000; Taniguchi et al. 2003); the nature of the membrane polymer backbone (Kaeselev et al. 2002; Kuroda et al. 1990; Yamagishi et al. 1995a, b; Puro et al. 2006); the UV intensity and the wavelength (Hwang and Park 2003; Lee et al. 2004; Qiu et al. 2005; Ji 1996); the employed additive(s) (Bhattacharya and Misra 2004); and the type of solvent(s) (Lee et al. 2004).

It is noted that the UV-initiated graft polymerization process is used predominantly for modification of ultrafiltration membranes (Kaeselev et al. 2002; Pieracci et al. 1999, 2000, 2002a, b; Qiu et al. 2005; Taniguchi et al. 2003; Yamagishi et al. 1995a, b; Susanto and Ulbricht 2008; Xi et al. 2006) and the hydrophilic monomer is usually utilised to enhance antifouling performance. Table 4.4 illustrates surface modification of PES via UV technique to enhance antifouling properties of membranes.

Gamma ray and electron beam-induced grafting

Gamma ray is rarely used as modification method for polymeric PES membrane because it consumes higher energy via the breaking of chemical bonds and therefore affects the membrane strength (Zhao et al. 2013a).

Electron beam irradiation, in contrast, is another approach to motivate grafting via generating active sites (for example, hydrophilic monomer) on the membrane surface. This technique is known to be a powerful method of modification used in the preparation of synthetic polymeric PES membranes with excellent properties including increased stability, improved separation efficiency and enhanced flux (Lai et al. 1986; Fritzsche et al. 1986; Haruvy et al. 1988; Saito et al. 1989). Electron beam irradiation technique can be used for preparation of anionic and cationic

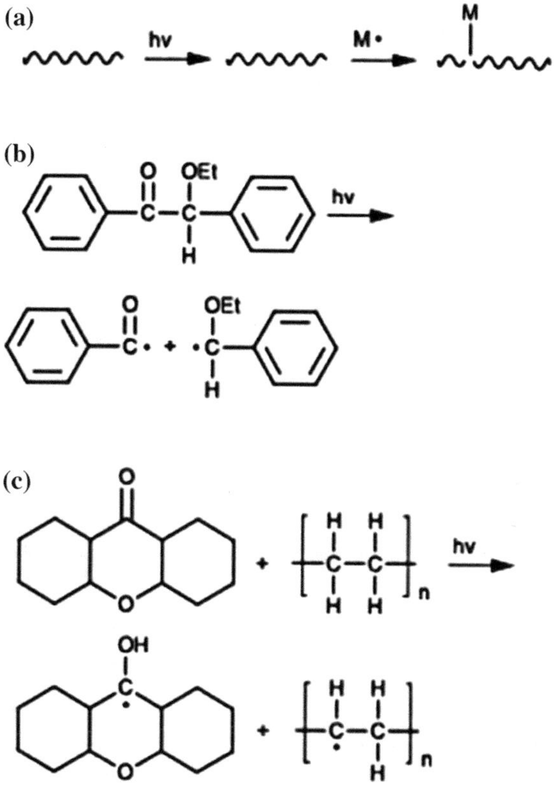

Fig. 4.4 A schematic diagram of the photoinitiated graft polymerization, **a** without a photoinitiator and **b** using a photoinitiator (adapted from Bhattacharya and Misra 2004)

exchange membranes. Keszler et al. (1991) have been modified PES membranes in a solution of AA and acryl amide monomers by using electron-induced grafting method. They concluded that the flux and solute retention were increased in comparison with the unmodified PES. Moreover, the PES membranes also had pH-responsivity due to the existence of hydroxyl groups (OH) in the PES network (backbone). Similar research studies conducted by Schulze et al. (2010), they have been modified PES membrane in a onestep process using electron beam method. The PES membranes were dipped in an aqueous solutions of functional molecules (e.g. sulfonic acid, carboxylic acid, phosphoric acid, amines, alcohols and zwitterionic compounds) followed by using electron beam treatment. The results showed that the protein adsorption of myoglobin and albumin was decreased significantly. There is no need to use any catalysts, photoinitiator, organic solvents, any toxic reagents and even additional synthetic or purification steps in this research study of Schulze et al. (2010). To sum up, electron beam irritation has the capability to interpenetrate to membrane polymer, and therefore, activate the internal surface of membrane for the desirable modification reaction.

Table 4.4 Summary of surface modification of polyethersulfone via UV technique

Membranes	Treatment	Function of membranes	References
PES (10 KDa)	Grafted with N-vinyl-2-pyrrolidinone (NVP)	UF: the membrane hydrophilicity was increased by 25 %, whilst the BSA rejection was decreased by 49 %	Pieracci et al. (1999)
PES (50 KDa)	Grafted with N-vinyl-2-pyrrolidinone (NVP)	UF: the treated PES membranes showed considerable decline of both membrane permeation and BSA retention	Pieracci et al. (2000)
PES	Acrylic acid irradiated with UV light	MF: the permeability of MF PES membrane was decreased because the modification process filled the pores with copolymer, FRR was 100 % for the treated PES membranes	Kouwonou et al. (2008)
PES	Grafted with AA through UV-initiated graft polymerization technique	NF: the membrane permeation was higher for the modified PES membrane in comparison with the neat PES membrane, the retention of humic acid was higher, the irreversible fouling was reduced by humic acid molecules, the pore size of membrane was decreased	Abu Seman et al. (2010)
PES	Grafted with hydrophilic functional moieties (viz. AA, AM) with UV-initiated photo-modification method	The treated PES membranes have a good separation capability in comparison to the neat PES, the flux recovery ratio (FRR) for the pristine PES membrane was minimum in comparison to other two grafted membranes, the antifouling performance of membrane was increased after grafting AA, increased in pH led to decrease separation performance	Gupta et al. (2009)
PES	Grafted with AMPS and qDMAEMA) using photoinduced grafting technique	MF: the number of bacterial cells was less for qDMAEMA-grafted samples in comparison to unmodified PES, higher anti-biofouling performance	Hilal et al. (2003)

(continued)

Table 4.4 (continued)

Membranes	Treatment	Function of membranes	References
PES	Heterogeneous Photograft copolymerization of PEGMA	UF: the modified membranes have higher resistance to fouling and higher retention, the membrane permeation was reduced due to pore blocking by grafted poly PEGMA	Susanto and Ulbricht (2007)
PES	Grafted with poly (ethylene glycol) methyl ether methacrylate (PEG) with vinylamides including *N*-vinylformamide (NVF), Nvinylacetamide (NVA) or *N*-methyl-*N*-vinylacetamide (MVA) via a photo-induced graft polymerization (PGP) method coupled with a high throughput platform (HTP) technology	The antifouling performance was improved significantly, the pore size of membrane as decreased	Yune et al. (2011)
PES	Immersion precipitation by TiO_2 nanoparticles and UV irradiation	UF: the initial permeate flux was lower than the initial permeate flux of the neat PES membrane	Rahimpour et al. (2008)

4.4.2 Plasma Treatment and Plasma-Induced Grafting Polymerisation

Over the last two decades, plasma treatment of the polymeric membranes has been extensively investigated by many researchers to enhance membrane hydrophilicity and to minimise membrane fouling to a minimum (Bryjak et al. 1999; Chen and Belfort 1999; Gancarz et al. 1999a, b, 2002, 2003; Ulbricht and Belfort 1996; Wavhal and Fisher 2002a; Zhan et al. 2004; Kim et al. 2002; Dattatray et al. 2005; Kull et al. 2005; Zhao et al. 2005; Dong et al. 2007; He et al. 2009; Poźniak et al. 2006; Tyszler et al. 2006; Yan et al. 2008; Yu et al. 2005, 2006, 2007, 2008a, b). It is an effective way to generate functional groups (radical sites that are stable in vacuum) on the polymeric membrane surface, leading to alterations in membrane performance. However, the bulk of the polymer remains unaffected. Because the lower surface energy of plasma treatment, the plasma is often restricted to the surface of membrane surface and therefore the bulk properties of membrane still remain unaffected. The thickness of the modified layer can be controlled up to the angstrom levels.

Kochkodan (2012) states that plasma treatment of the synthetic polymer membranes can be implemented in three various methods: (i) plasma treatment with non-polymerisable gas molecules; (ii) plasma treatment with polymerizable vapours; and (iii) plasma treatment with plasma-induced grafting. The plasma treatment with non-polymerizable gases and plasma-induced grafting are the most common methods used to modify the synthetic polymer membrane surfaces.

The advantages of plasma treatment are fast, effective technique and meet most of the ecological regulations for clean technology. It can be employed as radicals source, which work as anchor points for graft polymerisation.

A number of plasma techniques have been used to obtain permanent functional groups on the surface of the membrane. These techniques are: (1) Surface cleaning and etching (2) Surface modification with gas plasma (cross-linking and the creation of new functional groups) (3) Plasma-initiated polymerization/grafting (4) Plasma polymerization.

PES polymeric membranes can be modified through plasma, which is produced after ionisation of water vapour or a gas (e.g. Nitrogen, carbon dioxide, oxygen, hydrogen, argon) via an electrical discharge at elevated frequencies by using radio frequency waves and microwaves (Gancarz et al. 1999a, b; Van Der Bruggen 2009a; Castro Vidaurre et al. 2001; Zhao et al. 2013a).

Numerous gases can be employed for the modification of polymeric PES membrane (Saxena et al. 2009; Wavhal and Fisher 2002b, 2003), including Ar (Kull et al. 2005; Saxena et al. 2009; Wavhal and Fisher 2003), CO_2 (Pal et al. 2008; Wavhal and Fisher 2002b), N_2 (Kull et al. 2005), NH_3 (Iwa et al. 2004; Kull et al. 2005), O_2 (Pal et al. 2008; Saxena et al. 2009; Cho et al. 2004), and H_2O (Steen et al. 2002). The mechanism of their action is: the membrane surface is bombarded with ionised plasma components to produce radical sites C–C, C–H, and C–S bonds can then be attacked by radicals, with exclusion of the aromatic C–H and C–C bonds (Zhao et al. 2013a). This is the same to the photodegradation approach or process. The produced radicals can then react with gas molecules, creating different functionalities, depending on the conditions of plasma as shown in Fig. 4.5 for O_2 (Tyszler et al. 2006). The higher hydrophilicity of the treated membranes with oxygen plasma for a period of 20 s results in reducing liable membrane fouling to a minimum for ultrafiltration of gelatine solutions (Kim et al. 2002).

After contacting with the air, the remaining groups (radical sites) bind with nitrogen or oxygen. These functional groups are carbonyl, carboxylic and hydroxyl.

Surface modification of the polymeric membrane with CO_2 plasma results in introducing oxygen into the surface of membrane in the form of acid, carbonyl and ester groups, leading to an increase in membrane hydrophilicity. Generally, modification of membrane surfaces with CO_2 plasma results in hydrophilic surfaces and surface oxidation. He et al. (2009) conclude that treating membranes with CO_2 plasma for 30 s shows better regeneration behaviour in ultrafiltration membrane of specific filtrates (e.g. BSA protein solutions).

Membrane surface treatment with Nitrogen plasma leads to introduce different functional groups (e.g. imine, amine, amide, and nitrile) on the surface of membrane, making the membrane super hydrophilic and less liable/susceptible to fouling (Kull et al. 2005). Due to the weak effect of nitrogen plasma, the degradation of polymer may be reduced to certain extent.

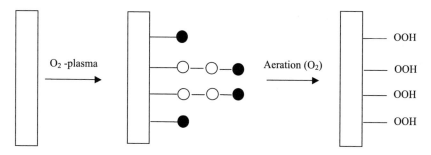

Fig. 4.5 Schematic representation of O₂ plasma treatment of a membrane (adapted from Tyszler et al. 2006)

Modification of polymeric membrane with H_2O plasma results in introduction of oxygen (contains functional groups) onto the membrane surface. XPS and FTIR can be used to characterise the functional groups. Membrane treatment with Ar plasma can allow PES membranes to graft or attach other polymers. Plasma treatment has advantages and disadvantages. The advantages are very shallow modification depth in comparison to other modification processes. The disadvantages are (1) time-dependent of the induced changes and therefore, it needs a vacuum system, which leads to an increase in operation cost, (2) chemical reactions of the plasma treatment are quite very complicated, so it is hard to understand the surface chemistry of the treated membrane in detail. Therefore, plasma treatment is not possible to apply on the large scale.

Thermally induced grafting and immobilisation

Thermal-induced grafting and immobilisation is simplistic routes to surface-modify PES membrane. Chemical initiators or cleavage agents are always used in this route. Moreover, many types of biomolecules, such as protein, enzyme and amino acid, could be covalently immobilised onto polymeric PES membranes through chemical reaction. It is hard to modify PES membranes by thermal-induced graft polymerization because the good stability of thermal-induced grafting.

Mu and Zhao (2009) modified of PES porous membranes using thermal-induced surface cross-linking process. Difunctional poly (ethylene glycol) diacrylate (PEGDA) was used as the main cross-linking modifier. Trifunctional trimethylol-propane trimethylacrylate (TMP TMA) was added into the reaction liquid (PEGDA solution) to enhance the rate of cross-linking on the surface of polymeric PES porous membranes. The membrane permeation and antifouling performance could be significantly improved when a moderate mass gain of approximately 150 mg/cm² was attained. The hydrophilicity of membrane was increased significantly due to the cross-linking process.

Another research study conducted by Mansourpanah et al. (2010), nanofiltration membranes were prepared by dissolving the blend of PES/Polyimide in DMF solvent, and Afterward the prepared NF membranes were modified using (PEGtriazine) as a new modifier material and methanol solution containing 10 wt%

ethylenediamine (EDA) as a cross-linker to open the rings of imide group. The membrane permeation and salt rejection was decreased and increased, respectively due to the addition of EDA and PEG-triazine. In case of incorporating trimethylacrylate into the membrane recipe, the modified membrane showed lower flux during the filtration process. Mansourpanah et al. (2010) stated that Diethanolamine could be also used as a hydrophilic modifier to alter membrane properties.

Shi et al. (2011) prepared poly (MA)-graft-polyethersulfone (PMAA-g-PES) membrane through non-solvent-induced phase separation process. The enzyme model trypsin (contains amino functional groups) was conjugated covalently to surface of membrane via 1-ethyl-(3-3-dimethylaminopropyl)-carbodiimide hydrochloride (EDC)/N-hydroxysuccinimide (NHS)-activation of carboxylic acid groups enriched on the membrane surface. The results showed that the treated membranes have desirable antifouling performance and superior self-cleaning properties. The modified membranes showed long-term life span for protein filtration due to the higher stability of trypsin-resided membrane surface. The existence of trypsin on the surface of membrane did not cause any obvious decline in membrane permeation. However, after filtration of protein solution, the membranes could recover their initial flux using washing process

Surface-initiated atom transfer radical polymerization (SI-ATRP)

In recent years, surface-initiated atom transfer radical polymerization (SI-ATRP) has been shown to one of the prevalent living/controlled polymerization method utilised for grafting polymers synthesis onto different or various surfaces in a controlled manner, including grafting density, chain length and chemical compositions, and therefore produce the high capacity of binding sites of functional molecules (Edmondson et al. 2004b; Liu et al. 2008; Xu et al. 2005; Zhao et al. 2013a). SI-ATRP method has gained popularity in membrane technology due to its unique properties, including regulated molecular weight, narrow molecular weight distribution; operate under hash or milder conditions, and tuneable decorations.

Until now, three kinds of grafting methods using ATRP to modify membranes have been demonstrated in the literature. These methods are grafting–to, grafting–through and grafting–from, which concern the synthesis and application of graft copolymers. The function of these methods is to modify membrane surfaces. Although all these methods are used, only a few journal articles have been published, regarding to use ATRP to modify PES membranes (Li et al. 2009a, b).

Grafting–to is rarely used as modification method of polymeric membranes. It is simpler than the grafting–from and it has not gained popularity due to the difficulty of reaching the high grafting density. Furthermore, the grafting density being restricted by space constraints around the active sites and the situation becomes worst. This leads to increases layer thickness and consequently the process becomes self-limiting (Edmondson et al. 2004a). In this technique, preformed polymer chains (carrying reactive anchor groups on the side chains or at the end) are attached covalently to the surface of membrane (Zhao and Brittain 2000). This technique is accurately controlled the structure of the grafted chain. UF PAN membranes have been functionalised by different low molecular-weight aromatic azide derivatives consists of various hydrophobic and hydrophilic functional groups (Ulbricht and Hicke 1993). The results showed that the protein-fouling intensity and separation properties were virtually changed, depending on the type of functional groups introduced. This was verified by increasing the membrane hydrophilicity and charged of the active membrane layer

Grafting–from method is commonly used for modification of various polymeric membranes (Kaeselev et al. 2002; Kilduff et al. 2000; Ma et al. 2000a, b; Pieracci et al. 1999, 2000, 2002; Ulbricht and Belfort 1996; Ulbricht et al. 1996; Yamagishi et al. 1995a, b; Yang and Yang 2003; Hilal et al. 2003, 2004; Taniguchi and Belfort 2004; Hu et al. 2006; Gu et al. 2009; Rahimpour et al. 2009; Yu et al. 2009a; Zhang et al. 2009; Abu Seman et al. 2010). In this technique, polymer chains directly grow in situ from the immobilised or grafted active groups (radicals) on the membrane surface in the existence monomers or initiators. The majority of these research studies have conducted on antifouling modification of the ultrafiltration and microfiltration membranes because higher grafting density is needed in order to use the grafted polymer layer as a selective barrier in reverse osmosis membrane (RO). Generally, this technique may be achieved with photoinitiator and without photoinitiator. Lastly, grafting-from has proven to be more feasible and versatile method than grafting-to.

The chloromethylation of PES under milder conditions lead to create surface benzyl chloride groups as the active sites for ATRP to adjust the hydrophobicity of the PES membrane (Zhao et al. 2013a). Functional hydrophilic polymer brushes of PEGMA and sodium 4-styrenesulfonate (NaStS), as well as their block copolymer brushes, were prepared through surface-initiated ATRP from the chloromethylated PES surfaces. PES membranes grafted with PEGMA membranes showed excellent protein-fouling resistance to protein adsorption (Li et al. 2009a).

Apart from ATRP, other polymerization methods, such as free radical mediated polymerization (SFRP), reversible addition-fragmentation chain transfer (RAFT) polymerisation, and iodine-transfer polymerization have been also used to modify of PES membranes.

Wang et al. (2014) modified PES UF membranes with halloysite nanotubes grafted with 2-methacryloyloxyethyl phosphorylcholine (HNTs-MPC) for the purpose of improving the antifouling performance of the membrane. HNTs have been chemically modified with MPC through reverse atom transfer radical polymerization RATRP. The resulting new hybrid membranes possessed higher pure water flux and good antifouling performance. The protein (BSA) adsorption experiment showed that the amounts of bovine serum albumin (BSA) adsorbed on the surface of membrane were decreased significantly. BSA ultrafiltration experiment also indicated that the hybrid PES membranes have higher antifouling property with the addition of HNTs-MPC than the virgin PES membrane. At the same time, the long-term ultrafiltration experiment showed that the hybrid PES membrane had an ideal stability (Wang et al. 2014).

Ozone-induced grafting of PES membranes

Due to the low concentration availability of ozone throughout the Earth's atmosphere, it has been used in a wide range of consumer and industrial applications. Liu et al. (2008) functionalised PES ultrafiltration membranes via grafted three hydrophilic functional polymers such as polyethylene glycol (PEG), polyvinyl alcohol (PVA), and CHI using UV/ozone methods. Based on their results, the surface modified PES membranes have higher surface roughness in comparison to the pristine PES membrane. It was also shown that the modified membranes have more hydrophilic surface and could reduce the amount of protein adsorption on membranes. This research studies have demonstrated that PEG is the most favourable result in modifying PES membranes for fouling mitigation among the three hydrophilic polymers tested. This has been verified by measurements of contact angle and protein adsorption experiment.

Redox-grafting PES membranes

Redox (oxidation-reduction reaction) is a method used in the 1980s for grafting of nylon and cellulose and later was used to polyamide membranes. Recently, it has been used to modify PES membranes (Belfer et al. 2000; Reddy et al. 2005). Redox grafting describes all chemical reactions in which atoms have their oxidation number (oxidation state) changed. Furthermore, redox grafting composed from two concepts: reduction and oxidation (Zhao et al. 2013a). Van Der Bruggen (2009b) states that redox initiators can be used to produce radicals under milder conditions with minimum side chain reaction. One of the benefits of redox reaction is the modification of the prepared membrane can be implemented in aqueous media at ambient temperature without oxygen removal.

Redox initiation occurs when the oxidant (e.g. persulphate) and the reductant (e.g. sulphite) exist in the reaction mixture. It can be seen that redox initiation are used for polymerizations, which occurs at temperature below 50 °C. The redox system is usually used $K_2S_2O_8/Na_2S_2O_5$, which have the capability to produce radicals on the surface of membrane for further grafting. $K_2S_2O_8$ represents the oxidants, whilst $K_2S_2O_5/Na_2S_2O_5$ represents the reductants. Redox-grafting reactions have been used for PA membranes. However, some research studies have been conducted on using redox reaction as method to modify PES membranes. For example, Reddy et al. (2005) have been modified PES membranes with acrylate monomers (AAm and methylene bisacryl amide) via graft polymerisation induced by a redox reaction.

Belfer et al. (2000) functionalised PES membranes using different monomers including AAm, MAA, PEGM and SPMA with the aid of redox initiators to create radicals on the PES membrane surface. They followed the same procedure used in the study of Reddy et al. (2005). They concluded that when a high concentration of monomer is used, the permeability might significantly decrease (up to 30 %) since the grafted chains can penetrate the membrane and block the membrane pores partially or completely.

Reddy et al. (2005) prepared PES nanofiltration membranes (MWCO < 1000) and polyamide composite membranes (thin film membrane contains negatively charged hydrophilic functional groups on its surface) using in situ redox polymerization of acrylate monomers. The resulting nanofiltration membranes were used for dye effluent solution treatment whilst the modified membranes were examined for desalination of brackish water without using chemical pre-treatment to evaluate their fouling resistance performance. Reddy et al. (2005) concluded that the treated membranes showed excellent antifouling properties for desalination of brackish water.

Pore-filled PES membranes

Pore-filled PES membranes are a new class of membranes developed by Mika et al. (1995). They provided pH-sensitive membrane, which consists of microfiltration (host) porous substrate and polyelectrolyte anchored within the pores. This method can also be used to modify PES membranes. The host substrate is porous and physically and chemically stable, provides mechanical strength for the developed membranes (Hu and Dickson 2007). Different polyelectrolyte can be pore filled in a controlled manner owing to the porous structure of the host substrate, resulting in dramatically different behaviour from the host membranes (Mika et al. 1995, 1999, 2002).

Shah et al. (2005) functionalized porous PES microfiltration membranes by the formation of polystyrene grafts within the pores. Further activation of the grafts is accomplished by lower concentration of sulfuric acid, resulting in no loss of functionalised polymer during activation method. The functionalised or treated membranes could also be employed for separation of whey protein (Cowan and Ritchie 2007). The surface charged membrane exhibited higher selectivity (five times) than the pristine PES membrane at pH 7.2. The selectivity was improved due to increase reduction of beta-lactoglobulin because a reduction in molecular sieving combined with electrostatic repulsion between the negatively charged membranes and negatively charged b-lactoglobulin.

Molecularly imprinted PES membranes

A molecular imprinting technique is a promising technique used to create highly recognition and catalytic sites within polymer network (Wulff 1995). Molecular imprinting technique can be prepared by copolymerisation of cross-linker and functional monomers into polymeric membrane matrix in the presence of a template molecule, which results in the formation of these monomers (binding or recognitions sites cavities) complementary to the template in shape, size and position of the functional groups (Mosbach and Haupt 1998). These recognition sites enable the polymer to rebind selectively the imprint molecules from a mixture of closely related compounds (Zhao et al. 2013a; Alexander et al. 2006; Cormack and Mosbach 1999; Davis et al. 1996; Mayes and Whitcombe 2005; Wulff 2002; Ye and Mosbach 2008; Wei et al. 2006a; Whitcombe and Vulfson 2001; Byrne et al. 2008). Due to its ease of operation, stability, higher selectivity, feasibility in different conditions, molecular imprinting technique has been extensively used in a wide range of applications, for example, chromatography, solid phase extraction, sensor design, adsorbents to environmental hormone, reaction catalysis and membrane separation (Bergmann and Peppas 2008; Byrne et al. 2002; Ge and Turner 2008; Hillberg and Tabrizian 2008; Hilt and Byrne 2004; Wei et al. 2006a; Bossi et al. 2007; Janiak and Kofinas 2007; Turner et al. 2006; Zhao et al. 2013a; Zhang et al. 2006). Molecular imprinting technique has gained popularity in industry due to its low cost and stability. Molecular imprinting technique has been developed by two main approaches: the first approach is covalent molecularly imprinting and the second one is non-covalent molecularly imprinting. The template is coupled to polymerizable molecules via covalent bond in the covalent approach, whilst, the template is mingled with functional monomers or cross-linker via non-covalent bond in non-covalent approach. The second approach has been used significantly due to its wide application and easiness. Byrne and Salian (2008) state that the majority of imprinted polymers generated today have been cross-linked extremely to minimise the flexibility of the associated binding cavities generated between polymer chains. Therefore, the idea of the technique translating to polymeric matrices with higher flexibility within their polymer chains was highly suspect. It is supposed that flexibility of polymeric chains would results in fatal deficiencies in the networks in which imprinted structures are well defined, namely template binding affinity, capacity, and selectivity. However, experimental work in the last decade has confirmed that this is not the case. Recently, Zhao et al. (2008) prepared bisphenol A imprinted PES hollow fiber membrane using phase transition technique and dry–wet-spinning method. The results showed that the transmission rate of bisphenol A has been increased for the imprinted hollow fiber membranes in comparison with the non-imprinted membranes because the larger amount of radical (binding) sites through the imprinted membranes. Therefore, the imprinted membranes can be used in separation and purification applications

Ionically modified PES membranes

Ionic modification is one of the methods used to modify polymeric PES membranes. Numerous research studies have been conducted to modify polymeric PES membranes ionically, for example, (Li and Chung 2010) modified PES membranes using ionic modification method to produce the sulfonated PES membranes with Ag^+ ion form. Silver ion modification approach and dual-layer hollow fiber spinning technology has been used to functionalise PES membranes. The results of pure gas permeance showed that the O_2/N_2 and CO_2/CH_4 selectivity was significantly increased (9.5 and 118), respectively. Cao et al. (2010) blended PES membranes with sulfonated PES to prepare hybrid membranes. The hybrid membranes were then immersed in a solution of excess $AgNO_3$. Silver ion were immobilised onto SPES membrane surface using Vitamin C (VC) as reducing agent. The results showed that the composite (PES/SPES)-Ag membranes have excellent bacteriostatic (growth inhibition) and anti-bacterial properties for long period of usage. In the future, the anti-bacterial properties of the composite (PES/SPES)-Ag membranes might be useful to extend the usage of PES in another applications such as food processing and medical instruments industry. Li et al. (2002) used co-extrusion and dry-jet wet-spinning phase inversion techniques to fabricate delamination-free-dual-layer asymmetric composite hollow fiber membranes with a defect-free dense selective skin. The modified membranes were used for applications of gas or air separation. The results of pure gas permeance showed that the O_2/ N_2 selectivity of the dual-layer asymmetric hollow fiber membranes at 4.6 was relatively close to the value of the outer-layer material, 4.7.

References

Abu Seman MN, Khayet M, Bin Ali ZI, Hilal N (2010) Reduction of nanofiltration membrane fouling by UV-initiated graft polymerization technique. J Membr Sci 355(1–2):133–141

Ajmani GS, Goodwin D, Marsh K, Fairbrother DH, Schwab KJ, Jacangelo JG, Huang H (2012) Modification of low pressure membranes with carbon nanotube layers for fouling control. Water Res 46(17):5645–5654

Akar N, Asar B, Dizge N, Koyuncu I (2013) Investigation of characterization and biofouling properties of PES membrane containing selenium and copper nanoparticles. J Membr Sci 437:216–226

Akhtar S, Hawes C, Dudley L, Reed I, Stratford P (1995) Coatings reduce the fouling of microfiltration membranes. J Membr Sci 107(3):209–218. doi:10.1016/0376-7388(95)00118-9

Akthakul A, Salinaro RF, Mayes AM (2004) Antifouling polymer membranes with subnanometer size selectivity. Macromolecules 37(20):7663–7668. doi:10.1021/ma048837s

Alexander C, Andersson HS, Andersson LI, Ansell RJ, Kirsch N, Nicholls IA, O'Mahony J, Whitcombe MJ (2006) Molecular imprinting science and technology: a survey of the literature for the years up to and including 2003. J Mol Recognit 19(2):106–180

Arthanareeswaran G, Thanikaivelan P, Raajenthiren M (2008) Fabrication and characterization of CA/PSf/SPEEK ternary blend ultrafiltration membranes. Ind Eng Chem Res 47(5):1488–1494

Bai P, Cao X, Zhang Y, Yin Z, Wei Q, Zhao C (2010) Modification of a polyethersulfone matrix by grafting functional groups and the research of biomedical performance. J Biomater Sci Polym Ed 21(12):1559–1572

Baker R (2012) Membrane technology and application, 3rd edn. Wiley, UK

Barzin J, Feng C, Khulbe KC, Matsuura T, Madaeni SS, Mirzadeh H (2004) Characterization of polyethersulfone hemodialysis membrane by ultrafiltration and atomic force microscopy. J Membr Sci 237(1–2):77–85

Batsch A, Tyszler D, Brügger A, Panglisch S, Melin T (2005) Foulant analysis of modified and unmodified membranes for water and wastewater treatment with LC-OCD. Desalination 178 (1–3 SPEC. ISS.):63–72

Belfer S, Fainchtain R, Purinson Y, Kedem O (2000) Surface characterization by FTIR-ATR spectroscopy of polyethersulfone membranes-unmodified, modified and protein fouled. J Membr Sci 172(1–2):113–124

Belfort G, Davis RH, Zydney AL (1994) The behavior of suspensions and macromolecular solutions in crossflow microfiltration. J Membr Sci 96(1–2):1–58

Bergmann NM, Peppas NA (2008) Molecularly imprinted polymers with specific recognition for macromolecules and proteins. Prog Polym Sci (Oxf) 33(3):271–288

Berrocal MJ, Daniel Johnson R, Badr IHA, Liu M, Gao D, Bachas LG (2002) Improving the blood compatibility of ion-selective electrodes by employing poly(MPC-co-BMA), a copolymer containing phosphorylcholine, as a membrane coating. Anal Chem 74(15):3644–3648. doi:10. 1021/ac025604v

Bhattacharya A, Misra BN (2004) Grafting: a versatile means to modify polymers: techniques, factors and applications. Prog Polym Sci (Oxf) 29(8):767–814

Blanco JF, Nguyen QT, Schaetzel P (2001) Novel hydrophilic membrane materials: sulfonated polyethersulfone Cardo. J Membr Sci 186(2):267–279. doi:10.1016/S0376-7388(01)00331-3

Boributh S, Chanachai A, Jiraratananon R (2009) Modification of PVDF membrane by chitosan solution for reducing protein fouling. J Membr Sci 342(1–2):97–104. doi:10.1016/j.memsci.2009.06.022

Bossi A, Bonini F, Turner APF, Piletsky SA (2007) Molecularly imprinted polymers for the recognition of proteins: the state of the art. Biosens Bioelectron 22(6):1131–1137

Bottino A, Capannelli G, Monticelli O, Piaggio P (2000) Poly(vinylidene fluoride) with improved functionalization for membrane production. J Membr Sci 166(1):23–29

Brink LES, Elbers SJG, Robbertsen T, Both P (1993) The anti-fouling action of polymers preadsorbed on ultrafiltration and microfiltration membranes. J Membr Sci 76(2–3):281–291

Bryjak M, Gancarz I, Poźniak G (1999) Surface evaluation of plasma-modified polysulfone (Udel P-1700) films. Langmuir 15(19):6400–6404

Byrne ME, Salian V (2008) Molecular imprinting within hydrogels II: progress and analysis of the field. Int J Pharm 364(2):188–212. doi:10.1016/j.ijpharm.2008.09.002

Byrne ME, Oral E, Hilt JZ, Peppas NA (2002) Networks for recognition of biomolecules: molecular imprinting and micropatterning poly (ethylene glycol)-containing films. Polym Adv Technol 13(10–12):798–816

Byrne M, Hilt J, Peppas N (2008) Recognitive biomimetic networks with moiety imprinting for intelligent drug delivery. Biomed Mater Res 84A:137–147

Cao X, Tang M, Liu F, Nie Y, Zhao C (2010) Immobilization of silver nanoparticles onto sulfonated polyethersulfone membranes as antibacterial materials. Colloids Surf B 81(2):555–562. doi:10.1016/j.colsurfb.2010.07.057

Castro Vidaurre EF, Achete CA, Simão RA, Habert AC (2001) Surface modification of porous polymeric membranes by RF-plasma treatment. Nucl Instrum Methods Phys Res, Sect B 175–177:732–736

Chae SR, Ahn YT, Kang ST, Shin HS (2006) Mitigated membrane fouling in a vertical submerged membrane bioreactor (VSMBR). J Membr Sci 280(1–2):572–581

Chen H, Belfort G (1999) Surface modification of poly(ether sulfone) ultrafiltration membranes by low-temperature plasma-induced graft polymerization. J Appl Polym Sci 72(13):1699–1711

Chen Y, Zhang Y, Zhang H, Liu J, Song C (2013) Biofouling control of halloysite nanotubes-decorated polyethersulfone ultrafiltration membrane modified with chitosan-silver nanoparticles. Chem Eng J 228:12–20

Cho D, Kim S, Huh Y, Kim D, Cho S, Kim B (2004) Effects of surface modification of the membrane in the ultrafiltation of waste water. Macromol Res 12:553–558

Chu LY, Wang S, Chen WM (2005) Surface modification of ceramic-supported polyethersulfone membranes by interfacial polymerization for reduced membrane fouling. Macromol Chem Phys 206(19):1934–1940. doi:10.1002/macp.200500324

Chu LQ, Tan WJ, Mao HQ, Knoll W (2006) Characterization of UV-induced graft polymerization of poly(acrylic acid) using optical waveguide spectroscopy. Macromolecules 39(25):8742–8746

Cormack PAG, Mosbach K (1999) Molecular imprinting: recent developments and the road ahead. React Funct Polym 41(1):115–124

Cowan S, Ritchie S (2007) Modified polyethersulfone (PES) ultrafiltration membranes for enhanced filtration of whey proteins. Sep Sci Technol 42(11):2405–2418. doi:10.1080/01496390701477212

Dang HT, Narbaitz RM, Matsuura T (2008) Double-pass casting: a novel technique for developing high performance ultrafiltration membranes. J Membr Sci 323(1):45–52

Daraei P, Madaeni SS, Ghaemi N, Khadivi MA, Astinchap B, Moradian R (2013a) Enhancing antifouling capability of PES membrane via mixing with various types of polymer modified multi-walled carbon nanotube. J Membr Sci 444:184–191

Daraei P, Madaeni SS, Ghaemi N, Khadivi MA, Astinchap B, Moradian R (2013b) Fouling resistant mixed matrix polyethersulfone membranes blended with magnetic nanoparticles: study of magnetic field induced casting. Sep Purif Technol 109:111–121

Dattatray S, Wavhal D, Fisher E (2005) Modification of polysulfone ultrafiltration membranes by CO_2 plasma treatment. Desalination 172:198–205

Davis ME, Katz A, Ahmad WR (1996) Rational catalyst design via imprinted nanostructured materials. Chem Mater 8(8):1820–1839

Deng B, Yang X, Xie L, Li J, Hou Z, Yao S, Liang G, Sheng K, Huang Q (2009) Microfiltration membranes with pH dependent property prepared from poly(methacrylic acid) grafted polyethersulfone powder. J Membr Sci 330(1–2):363–368

Devrim Y, et al (2009) Preparation and characterization of sulfonated polysulfone/titanium dioxide composite membranes for proton exchange membrane fuel cells. Int J Hydrogen Energ 34 (8):3467–3475

Dong B, Jiang H, Manolache S, Wong ACL, Denes FS (2007) Plasma-mediated grafting of poly (ethylene glycol) on polyamide and polyester surfaces and evaluation of antifouling ability of modified substrates. Langmuir 23(13):7306–7313

Dong C, He G, Li H, Zhao R, Han Y, Deng Y (2012) Antifouling enhancement of poly(vinylidene fluoride) microfiltration membrane by adding Mg(OH) 2 nanoparticles. J Membr Sci 387–388 (1):40–47

Du JR, Peldszus S, Huck PM, Feng X (2009) Modification of poly(vinylidene fluoride) ultrafiltration membranes with poly(vinyl alcohol) for fouling control in drinking water treatment. Water Res 43(18):4559–4568. doi:10.1016/j.watres.2009.08.008

Duan L, Huang W, Zhang Y (2015) High-flux, antibacterial ultrafiltration membranes by facile blending with N-halamine grafted halloysite nanotubes. RSC Adv 5(9):6666–6674. doi:10. 1039/c4ra14530e

Edmondson S, Osborne V, Huck W (2004a) Chemical Soc Rev 35:14

Edmondson S, Osborne VL, Huck WTS (2004b) Polymer brushes via surface-initiated polymerizations. Chem Soc Rev 33(1):14–22

Emrick TS, Breitenkamp K, Revanur R, Freeman BD, McCloskey (2008) B. Chem. Abstr 149, 54430; U.S. Patent Appl. 2008/0142454

Fan Z, Wang Z, Sun N, Wang J, Wang S (2008) Performance improvement of polysulfone ultrafiltration membrane by blending with polyaniline nanofibers. J Membr Sci 320(1–2): 363–371

Fang B, Ling Q, Zhao W, Ma Y, Bai P, Wei Q, Li H, Zhao C (2009) Modification of polyethersulfone membrane by grafting bovine serum albumin on the surface of polyethersulfone/poly(acrylonitrile-co-acrylic acid) blended membrane. J Membr Sci 329(1–2):46–55

Fang BH, Cheng C, Li LL, Cheng J, Zhao WF, Zhao CS (2010) Surface modification of polyethersulfone membrane by grafting bovine serum albumin. Fibers and Polymers 11 (7):960–966

Filho AAMF, Gomes AS (2006) Copolymerization of styrene onto polyethersulfone films induced by gamma ray irradiation. Polym Bull 57:415–21

Freeman, BD Pinnau I (2004) In Advanced Materials for Membrane Separations; Pinnau I, Freeman BD, (eds) ACS Symp Ser 876; American

Fritzsche AK (1986) Effect of ionizing radiation on styrene/acrylonitrile copolymer hollow fiber membranes. J Appl Polym Sci 32(2):3541–3550

Galjaard G, Buijs P, Beerendonk E, Schoonenberg F, Schippers JÇ (2001) Pre-coating (EPCE®) UF membranes for direct treatment of surface water. Desalination 139(1–3):305–316. doi:10. 1016/S0011-9164(01)00324-1

Gancarz I, Poźniak G, Bryjak M (1999a) Modification of polysulfone membranes 1. CO_2 plasma treatment. Eur Polymer J 35(8):1419–1428

Gancarz I, Pozniak J, Bryjak M, Frankiewicz A (1999b) Modification of polysulfone membranes. 2. Plasma grafting and plasma polymerization of acrylic acid. Acta Polymerica 50 (9):317–326

Gancarz I, Poźniak G, Bryjak M, Tylus W (2002) Modification of polysulfone membranes 5. Effect of n-butylamine and allylamine plasma. Eur Polymer J 38(10):1937–1946

Gancarz I, Bryjak J, Bryjak M, Poźniak G, Tylus WLS (2003) Plasma modified polymers as a support for enzyme immobilization 1. Allyl alcohol plasma. Eur Polym J 39(8):1615–1622

Ge Y, Turner APF (2008) Too large to fit? Recent developments in macromolecular imprinting. Trends Biotechnol 26(4):218–224

Gu JS, Yu HY, Huang L, Tang ZQ, Li W, Zhou J, Yan MG, Wei XW (2009) Chain-length dependence of the antifouling characteristics of the glycopolymer-modified polypropylene membrane in an SMBR. J Membr Sci 326(1):145–152

Gupta BB, Blanpain P, Jaffrin MY (1992) Permeate flux enhancement by pressure and flow pulsations in microfiltration with mineral membranes. J Membr Sci 70(2–3):257–266. doi:10. 1016/0376-7388(92)80111-V

Gupta S, Yogesh Singh K, Bhattacharya A (2009) Studies on permeation of Bovine Serum Albumin (BSA) through photo-modified functionalized asymmetric membrane. J Macromol Sci Part A Pure Appl Chem 46(1):90–96. doi:10.1080/10601320802515506

Ha HY, Nam SW, Lim TH, Oh IH, Hong SA (1996) Properties of the TiO_2 membranes prepared by CVD of titanium tetraisopropoxide. J Membr Sci 111(1):81–92

Hai FI, Yamamoto K, Fukushi K, Nakajima F (2008) Fouling resistant compact hollow-fiber module with spacer for submerged membrane bioreactor treating high strength industrial wastewater. J Membr Sci 317(1–2):34–42

Haruvy Y, Gratzel M, RaJbenbach LA (1988) Radlolytlc method of preparation of semiconductor assemblies supported on polymenc membranes. Radlat Phys Chem 31:843–852

He XC, Yu HY, Tang ZQ, Liu LQ, Yan MG, Gu JS, Wei XW (2009) Reducing protein fouling of a polypropylene microporous membrane by CO_2 plasma surface modification. Desalination 244(1–3):80–89

Higuchi A, Sugiyama K, Yoon BO, Sakurai M, Hara M, Sumita M, Sugawara SI, Shirai T (2003) Serum protein adsorption and platelet adhesion on pluronic™-adsorbed polysulfone membranes. Biomaterials 24(19):3235–3245. doi:10.1016/S0142-9612(03)00186-8

Hilal N, Al-Khatib L, Atkin BP, Kochkodan V, Potapchenko N (2003) Photochemical modification of membrane surfaces for (bio)fouling reduction: a nano-scale study using AFM. Desalination 158(1–3):65–72

Hilal N, Kochkodan V, Al-Khatib L, Levadna T (2004) Surface modified polymeric membranes to reduce (bio)fouling: a microbiological study using *E. coli*. Desalination 167(1–3):293–300

Hilal N, Ogunbiyi OO, Miles NJ, Nigmatullin R (2005) Methods employed for control of fouling in MF and UF membranes: a comprehensive review. Sep Sci Technol 40(10):1957–2005

Hillberg AL, Tabrizian M (2008) Biomolecule imprinting: developments in mimicking dynamic natural recognition systems. ITBM-RBM 29(2–3):89–104

Hilt JZ, Byrne ME (2004) Configurational biomimesis in drug delivery: Molecular imprinting of biologically significant molecules. Adv Drug Deliv Rev 56(11):1599–1620

Hu K, Dickson JM (2007) Development and characterization of poly(vinylidene fluoride)-poly (acrylic acid) pore-filled pH-sensitive membranes. J Membr Sci 301(1–2):19–28

Hu M, Yang Q, Xu Z (2006) Enhancing the hydrophilicity of polyproplylene microporous membranes by the grafting of 2-hydroxyl metharylate via synergistic effect of photoinitiators. Membrane Science 285:196–205

Huang J, Zhang K, Wang K, Xie Z, Ladewig B, Wang H (2012) Fabrication of polyethersulfone-mesoporous silica nanocomposite ultrafiltration membranes with antifouling properties. J Membr Sci 423–424:362–370

Hvid KB, Nielsen PS, Stengaard FF (1990) Preparation and characterization of a new ultrafiltration membrane. J Membr Sci 53(3):189–202. doi:10.1016/0376-7388(90)80014-D

Hwang TS, Park JW (2003) UV-induced graft polymerization of polypropylene-g-glycidyl methacrylate membrane in the vapor phase. Macromol Res 11(6):495–500

Iojoiu C, Maréchal M, Chabert F, Sanchez JY (2005) Mastering sulfonation of aromatic polysulfones: crucial for membranes for fuel cell application. Fuel Cells 5(3):344–354. doi:10. 1002/fuce.200400082

Iwa T, Kumazawa H, Bae SY (2004) Gas permeabilities of NH3-plasma-treated polyethersulfone membranes. J Appl Polym Sci 94(2):758–762

Jamshidi Gohari R, Halakoo E, Nazri NAM, Lau WJ, Matsuura T, Ismail AF (2014) Improving performance and antifouling capability of PES UF membranes via blending with highly hydrophilic hydrous manganese dioxide nanoparticles. Desalination 335(1):87–95

Janiak DS, Kofinas P (2007) Molecular imprinting of peptides and proteins in aqueous media. Anal Bioanal Chem 389(2):399–404

Ji J (1996) Fabrication and photochemical surface modification of photoreactive thin-film composite membranes and model development for thin film formation by interfacial polymerization. McMaster University

Kaeselev B, Pieracci J, Belfort G (2001) Photoinduced grafting of ultrafiltration membranes: comparison of poly(ether sulfone) and poly(sulfone). J Membr Sci 194(2):245–261. doi:10.1016/S0376-7388(01)00544-0

Kaeselev B, Kingshott P, Jonsson G (2002) Influence of the surface structure on the filtration performance of UV-modified PES membranes. Desalination 146(1–3):265–271

Kang G, Liu M, Lin B, Cao Y, Yuan Q (2007) A novel method of surface modification on thin-film composite reverse osmosis membrane by grafting poly(ethylene glycol). Polymer 48 (5):1165–1170

Kang GD, Cao YM (2014) Application and modification of poly (vinylidene fluoride) (PVDF) membranes—A review. J Membrane Sci 463:145–165

Keszler B, Kovács G, Tóth A, Bertóti I, Hegyi M (1991) Modified polyethersulfone membranes. J Membr Sci 62(2):201–210

Khayet M, Suk DE, Narbaitz RM, Santerre JP, Matsuura T (2003) Study on surface modification by surface-modifying macromolecules and its applications in membrane-separation processes. J Appl Polym Sci 89(11):2902–2916. doi:10.1002/app.12231

Khulbe KC, Feng C, Matsuura T (2010a) The art of surface modification of synthetic polymeric membranes. J Appl Polym Sci 115(2):855–895

Khulbe KC, Feng CY, Matsuura T (2010b) Surface modification of synthetic polymeric membranes for filtration and gas separation. Recent Pat Chem Eng 3(1):1–16

Kilduff JE, Mattaraj S, Pieracci JP, Belfort G (2000) Photochemical modification of poly(ether sulfone) and sulfonated poly(sulfone) nanofiltration membranes for control of fouling by natural organic matter. Desalination 132(1–3):133–142

Kim KS, Lee KH, Cho K, Park CE (2002) Surface modification of polysulfone ultrafiltration membrane by oxygen plasma treatment. J Membr Sci 199(1):135–145

Kochkodan V, Hilal N (2015) A comprehensive review on surface modified polymer membranes for biofouling mitigation. Desalination 356:187–207. doi:10.1016/j.desal.2014.09.015

Kouwonou Y, Malaisamy R, Jones KL (2008) Modification of PES membrane: reduction of biofouling and improved flux recovery. Sep Sci Technol 43(16):4099–4112. doi:10.1080/01496390802414726

Kroll S, Meyer L, Graf AM, Beutel S, Glökler J, Döring S, Klaus U, Scheper T (2007) Heterogeneous surface modification of hollow fiber membranes for use in micro-reactor systems. J Membr Sci 299(1–2):181–189

Kull KR, Steen ML, Fisher ER (2005) Surface modification with nitrogen-containing plasmas to produce hydrophilic, low-fouling membranes. J Membr Sci 246(2):203–215

Kuroda S, Mita I, Obata K, Tanaka S (1990) Degradation of aromatic polymers: part IV-Effect of temperature and light intensity on the photodegradation of polyethersulfone. Polym Degrad Stab 27(3):257–270

La YH, McCloskey BD, Sooriyakumaran R, Vora A, Freeman B, Nassar M, Hedrick J, Nelson A, Allen R (2011) Bifunctional hydrogel coatings for water purification membranes: improved fouling resistance and antimicrobial activity, J Membr Sci 372:285–291

Lai JY, Chang TC, Wu ZJ, Hsieh TS (1986) Vinyl monomer irradiation-grafted nylon 4 membranes. J Appl Polym Sci 32(4):4709–4718

Le-Clech P, Cao Z, Wan PY, Wiley DE, Fane AG (2006a) The application of constant temperature anemometry to membrane processes. J Membr Sci 284(1–2):416–423

Le-Clech P, Chen V, Fane TAG (2006b) Fouling in membrane bioreactors used in wastewater treatment. J Membr Sci 284(1–2):17–53

Le-Clech P, Marselina Y, Stuetz R, Chen V (2006c) Fouling visualisation of soluble microbial product models in MBRs. Desalination 199(1–3):477–479

Lee D, Kim H, Kim S (2004) Surface modification of polymeric membranes by UV grafting. In: Separation Amfm (ed) ACS Symposium Series 876. American Chemical Society, Washington, pp 281–299

Li Y, Chung TS (2010) Silver ionic modification in dual-layer hollow fiber membranes with significant enhancement in CO_2/CH_4 and O_2/N_2 separation. J Membr Sci 350(1–2):226–231. doi:10.1016/j.memsci.2009.12.032

Li DF, Chung TS, Wang R, Liu Y (2002) Fabrication of fluoropolyimide/polyethersulfone (PES) dual-layer asymmetric hollow fiber membranes for gas separation. J Membr Sci 198 (2):211–223

Li L, Yan G, Wu J (2009a) Modification of polysulfone membranes via surface initiated atom transfer radical polymerization and their antifouling properties Applied polymer science 111:1942–1946

Li SD, Wang CC, Chen CY (2009a) Water electrolysis for H2 production using a novel bipolar membrane in low salt concentration. J Membr Sci 330(1–2):334–340

Li L, Yan G, Wu J, Yu X, Guo Q (2009b) Surface-initiated atom-transfer radical polymerization from polyethersulfone membranes and their antifouling properties. e-Polymers, 26

Li L, Yan G, Wu J, Yu X, Guo Q (2009c) Surface-initiated atom-transfer radical polymerization from polyethersulfone membranes and their use in antifouling. e-Polymers, 1–10

Li X, Kita H, Zhu H, Zhang Z, Tanaka K (2009e) Synthesis of long-term acid-stable zeolite membranes and their potential application to esterification reactions. J Membr Sci 339(1–2):224–232

Li ZL, Zhang YZ, Wang XJ, Li HB (2009f) Pervaporation and its applications in dehydration of hydrazine fuels. Hanneng Cailiao/Chinese J Energ Mater 17(1):107–112

Li L, Yin Z, Li F, Xiang T, Chen Y, Zhao C (2010) Preparation and characterization of poly (acrylonitrile-acrylic acid-N-vinyl pyrrolidinone) terpolymer blended polyethersulfone membranes. J Membr Sci 349(1–2):56–64

Li F, Meng J, Ye J, Yang B, Tian Q, Deng C (2014) Surface modification of PES ultrafiltration membrane by polydopamine coating and poly(ethylene glycol) grafting: morphology, stability, and anti-fouling. Desalination 344:422–430

Liang L, Rieke PC, Fryxell GE, Liu J, Engehard MH, Alford KL (2000) Temperature-sensitive surfaces prepared by UV photografting reaction of photosensitizer and N-Isopropylacrylamide. J Phys Chem B 104(49):11667–11673

Lin YF, Tung KL, Tzeng YS, Chen JH, Chang KS (2012) Rapid atmospheric plasma spray coating preparation and photocatalytic activity of macroporous titania nanocrystalline membranes. J Membr Sci 389:83–90

Liu SX, Kim JT (2011) Characterization of surface modification of polyethersulfone membrane. J Adhes Sci Technol 25(1–3):193–212

Liu SX, Kim JT, Kim S (2008) Effect of polymer surface modification on polymer-protein interaction via hydrophilic polymer grafting. J Food Sci 73(3):E143–E150

Liu Z, Deng X, Wang M, Chen J, Zhang A, Gu Z, Zhao C (2009) BSA-modified polyethersulfone membrane: preparation, characterization and biocompatibility. J Biomater Sci Polym Ed 20 (3):377–397

Luo ML, Zhao JQ, Tang W, Pu CS (2005) Hydrophilic modification of poly(ether sulfone) ultrafiltration membrane surface by self-assembly of TiO_2 nanoparticles. Appl Surf Sci 249(1–4):76–84

Luo M, Wen Q, Liu J, Liu H, Jia Z (2011) Fabrication of SPES/nano-TiO_2 composite ultrafiltration membrane and its anti-fouling mechanism. Chin J Chem Eng 19(1):45–51

Ma H, Bowman CN, Davis RH (2000a) Membrane fouling reduction by backpulsing and surface modification. J Membr Sci 173(2):191–200

Ma SD, Sun HY, Huang G, Zhu X (2000b) Effect of marine fouling creatures on corrosion of carbon steel. J Chin Soc Corros Prot 20(3):181–182

Ma W, Zhang J, Wang X, Wang S (2007) Effect of PMMA on crystallization behavior and hydrophilicity of poly(vinylidene fluoride)/poly(methyl methacrylate) blend prepared in semi-dilute solutions. Appl Surf Sci 253(20):8377–8388

Maartens A, Jacobs EP, Swart P (2002) UF of pulp and paper effluent: membrane fouling-prevention and cleaning. J Membr Sci 209(1):81–92. doi:10.1016/S0376-7388(02) 00266-1

Madaeni SS, Zinadini S, Vatanpour V (2011) A new approach to improve antifouling property of PVDF membrane using in situ polymerization of PAA functionalized TiO_2 nanoparticles. J Membr Sci 380(1–2):155–162

Majeed S, Fierro D, Buhr K, Wind J, Du B, Boschetti-de-Fierro A, Abetz V (2012) Multi-walled carbon nanotubes (MWCNTs) mixed polyacrylonitrile (PAN) ultrafiltration membranes. J Membr Sci 403–404:101–109

Malaisamy R, Berry D, Holder D, Raskin L, Lepak L, Jones KL (2010) Development of reactive thin film polymer brush membranes to prevent biofouling. J Membr Sci 350(1–2):361–370. doi:10.1016/j.memsci.2010.01.011

Mansourpanah Y, Madaeni SS, Adeli M, Rahimpour A, Farhadian A (2009) Surface modification and preparation of nanofiltration membrane from polyethersulfone/polyimide blend-use of a new material (polyethyleneglycol-triazine). J Appl Polym Sci 112(5):2888–2895. doi:10.1002/app.29821

Mansourpanah Y, Madaeni SS, Rahimpour A, Kheirollahi Z, Adeli M (2010) Changing the performance and morphology of polyethersulfone/polyimide blend nanofiltration membranes using trimethylamine. Desalination 256(1–3):101–107. doi:10.1016/j.desal.2010.02.006

Martín A, Arsuaga JM, Roldán N, de Abajo J, Martínez A, Sotto A (2015) Enhanced ultrafiltration PES membranes doped with mesostructured functionalized silica particles. Desalination 357:16–25. doi:10.1016/j.desal.2014.10.046

Mayes AG, Whitcombe MJ (2005) Synthetic strategies for the generation of molecularly imprinted organic polymers. Adv Drug Deliv Rev 57(12):1742–1778

McCloskey BD, Park HB, Ju H, Rowe BW, Miller DJ, Freeman BD (2012) A bioinspired fouling-resistant surface modification for water purification membranes. J Membr Sci 413–414:82–90

Mika AM, Childs RF, Dickson JM, McCarry BE, Gagnon DR (1995) A new class of polyelectrolyte-filled microfiltration membranes with environmentally controlled porosity. J Membr Sci 108(1–2):37–56

Mika AM, Childs RF, Dickson JM (1999) Chemical valves based on poly(4-vinylpyridine)-filled microporous membranes. J Membr Sci 153(1):45–56

Mika AM, Childs RF, Dickson JM (2002) Salt separation and hydrodynamic permeability of a porous membrane filled with pH-sensitive gel. J Membr Sci 206(1–2):19–30

Millesime L, Amiel C, Chaufer B (1994a) Ultrafiltration of lysozyme and bovine serum albumin with polysulfone membranes modified with quaternized polyvinylimidazole. J Membr Sci 89 (3):223–234. doi:10.1016/0376-7388(94)80104-5

Millesime L, Amiel C, Chaufer B (1994b) Ultrafiltration of lysozyme and bovine serum albumin with polysulfone membranes modified with quaternized polyvinylimidazole. J Membrane Sci Journal of membrane Science 89:223–234

Mishra SP, Vijaya (2007) Removal behavior of hydrous manganese oxide and hydrous stannic oxide for Cs(I) ions from aqueous solutions. Sep Purif Technol 54(1):10–17

Mittal KL (2009) Surface modification techniques. In: Polymer Surface Modification. CRC Press, Leiden

Moghimifar V, Raisi A, Aroujalian A (2014) Surface modification of polyethersulfone ultrafiltration membranes by corona plasma-assisted coating TiO2 nanoparticles. J Membr Sci 461:69–80

Mohd Yusof AH, Ulbricht M (2008) Polypropylene-based membrane adsorbers via photo-initiated graft copolymerization: optimizing separation performance by preparation conditions. J Membr Sci 311(1–2):294–305

Mosbach K, Haupt K (1998) Some new developments and challenges in noncovalent molecular imprinting technology. J Mol Recognit 11(1–6):62–68

Mu LJ, Zhao WZ (2009) Hydrophilic modification of polyethersulfone porous membranes via a thermal-induced surface crosslinking approach. Appl Surf Sci 255(16):7273–7278. doi:10. 1016/j.apsusc.2009.03.081

Nair AK, Isloor AM, Kumar R, Ismail AF (2013) Antifouling and performance enhancement of polysulfone ultrafiltration membranes using CaCO$_3$ nanoparticles. Desalination 322:69–75

Nunes SP, Sforça ML, Peinemann KV (1995) Dense hydrophilic composite membranes for ultrafiltration. J Membr Sci 106(1–2):49–56. doi:10.1016/0376-7388(95)00076-O

Nyström M (1989) Fouling of unmodified and modified polysulfone ultrafiltration membranes by ovalbumin. J Membr Sci 44(2–3):183–196. doi:10.1016/S0376-7388(00)83351-7

Pal S, Ghatak SK, De S, DasGupta S (2008) Evaluation of surface roughness of a plasma treated polymeric membrane by wavelet analysis and quantification of its enhanced performance. Appl Surf Sci 255(5 PART 1):2504–2511

Pang R, Li X, Li J, Lu Z, Sun X, Wang L (2014) Preparation and characterization of ZrO$_2$/PES hybrid ultrafiltration membrane with uniform ZrO$_2$ nanoparticles. Desalination 332:60–66. doi:10.1016/j.desal.2013.10.024

Parida KM, Kanungo SB, Sant BR (1981) Studies on MnO$_2$-I. Chemical composition, microstructure and other characteristics of some synthetic MnO$_2$ of various crystalline modifications. Electrochim Acta 26(3):435–443

Park JY, Acar MH, Akthakul A, Kuhlman W, Mayes AM (2006) Polysulfone-graft-poly(ethylene glycol) graft copolymers for surface modification of polysulfone membranes. Biomaterials 27 (6):856–865

Peeva PD, Million N, Ulbricht M (2012) Factors affecting the sieving behavior of anti-fouling thin-layer cross-linked hydrogel polyethersulfone composite ultrafiltration membranes. J Membr Sci 390–391:99–112

Peyravi M, Rahimpour A, Jahanshahi M (2012) Thin film composite membranes with modified polysulfone supports for organic solvent nanofiltration. J Membr Sci 423–424:225–237

Pieracci J, Crivello JV, Belfort G (1999) Photochemical modification of 10 kDa polyethersulfone ultrafiltration membranes for reduction of biofouling. J Membr Sci 156(2):223–240

Pieracci J, Wood DW, Crivello JV, Belfort G (2000) UV-assisted graft polymerization of N-vinyl-2-pyrrolidinone onto poly(ether sulfone) ultrafiltration membranes: Comparison of dip versus immersion modification techniques. Chem Mater 12(8):2123–2133

Pieracci J, Crivello JV, Belfort G (2002a) Increasing membrane permeability of UV-modified poly (ether sulfone) ultrafiltration membranes. J Membr Sci 202(1–2):1–16

Pieracci J, Crivello JV, Belfort G (2002b) UV-assisted graft polymerization of N-vinyl-2-pyrrolidinone conto poly(ether sulfone) ultrafiltration membranes using selective UV wavelengths. Chem Mater 14(1):256–265

Pinnau I, Freeman BD (2000) Formation and modification of polymeric membranes: overview. ACS symposium

Poźniak G, Gancarz I, Tylus W (2006) Modified poly(phenylene oxide) membranes in ultrafiltration and micellar-enhanced ultrafiltration of organic compounds. Desalination 198 (1–3):215–224

Prihandana GS, Ito H, Nishinaka Y, Kanno Y, Miki N (2012) Polyethersulfone membrane coated with nanoporous parylene for ultrafiltration. J Microelectromech Syst 21(6):1288–1290. doi:10.1109/JMEMS.2012.2224643

Puro L, Mänttäri M, Pihlajamäki A, Nyström M (2006) Characterization of modified nanofiltration membranes by octanoic acid permeation and FTIR analysis. Chem Eng Res Des 84(2A):87–96

Qiu C, Xu F, Nguyen QT, Ping Z (2005) Nanofiltration membrane prepared from cardo polyetherketone ultrafiltration membrane by UV-induced grafting method. J Membr Sci 255(1–2):107–115

Qiu C, Nguyen QT, Ping Z (2007) Surface modification of cardo polyetherketone ultrafiltration membrane by photo-grafted copolymers to obtain nanofiltration membranes. J Membr Sci 295 (1–2):88–94

Rahimpour A (2011a) Preparation and modification of nano-porous polyimide (PI) membranes by UV photo-grafting process: Ultrafiltration and nanofiltration performance. Korean J Chem Eng 28(1):261–266

Rahimpour A (2011b) UV photo-grafting of hydrophilic monomers onto the surface of nano-porous PES membranes for improving surface properties. Desalination 265(1–3):93–101

Rahimpour A, Madaeni SS (2010) Improvement of performance and surface properties of nano-porous polyethersulfone (PES) membrane using hydrophilic monomers as additives in the casting solution. J Membr Sci 360(1–2):371–379

Rahimpour A, Madaeni SS, Taheri AH, Mansourpanah Y (2008) Coupling TiO_2 nanoparticles with UV irradiation for modification of polyethersulfone ultrafiltration membranes. J Membr Sci 313(1–2):158–169

Rahimpour A, Madaeni SS, Zereshki S, Mansourpanah Y (2009) Preparation and characterization of modified nano-porous PVDF membrane with high antifouling property using UV photo-grafting. Appl Surf Sci 255(16):7455–7461

Rahimpour A, Madaeni SS, Ghorbani S, Shockravi A, Mansourpanah Y (2010a) The influence of sulfonated polyethersulfone (SPES) on surface nano-morphology and performance of polyethersulfone (PES) membrane. Appl Surf Sci 256(6):1825–1831. doi:10.1016/j.apsusc. 2009.10.014

Rahimpour A, Madaeni SS, Mansourpanah Y (2010b) Fabrication of polyethersulfone (PES) membranes with nano-porous surface using potassium perchlorate (KClO4) as an additive in the casting solution. Desalination 258(1–3):79–86

Rahimpour A, Madaeni SS, Mansourpanah Y (2010c) Nano-porous polyethersulfone (PES) membranes modified by acrylic acid (AA) and 2-hydroxyethylmethacrylate (HEMA) as additives in the gelation media. J Membr Sci 364(1–2):380–388

Rajagopalan M, Ramamoorthy M, Doraiswamy RM (2004) Cellulose acetate and polyethersulfone blend ultrafiltration membranes. Part I. Preparation and characterizations. Polym Adv Technol 15:149–157

Ran F, Nie S, Zhao W, Li J, Su B, Sun S, Zhao C (2011) Biocompatibility of modified polyethersulfone membranes by blending an amphiphilic triblock co-polymer of poly(vinyl pyrrolidone)-b-poly(methyl methacrylate)-b-poly(vinyl pyrrolidone). Acta Biomater 7 (9):3370–3381

Rana D, Matsuura T (2010) Surface modifications for antifouling membranes. Chem Rev 110 (4):2448–2471

Razali NF, Mohammad AW, Hilal N, Leo CP, Alam J (2013) Optimisation of polyethersulfone/polyaniline blended membranes using response surface methodology approach. Desalination 311:182–191

Razmjou A, Mansouri J, Chen V (2011a) The effects of mechanical and chemical modification of TiO_2 nanoparticles on the surface chemistry, structure and fouling performance of PES ultrafiltration membranes. J Membr Sci 378(1–2):73–84

Razmjou A, Mansouri J, Chen V, Lim M, Amal R (2011b) Titania nanocomposite polyether-sulfone ultrafiltration membranes fabricated using a low temperature hydrothermal coating process. J Membr Sci 380(1–2):98–113

Razmjou A, Resosudarmo A, Holmes RL, Li H, Mansouri J, Chen V (2012) The effect of modified TiO_2 nanoparticles on the polyethersulfone ultrafiltration hollow fiber membranes. Desalination 287:271–280

Reddy AVR, Mohan DJ, Bhattacharya A, Shah VJ, Ghosh PK (2003) Surface modification of ultrafiltration membranes by preadsorption of a negatively charged polymer: I. Permeation of water soluble polymers and inorganic salt solutions and fouling resistance properties. J Membr Sci 214(2):211–221

Reddy AVR, Trivedi JJ, Devmurari CV, Mohan DJ, Singh P, Rao AP, Joshi SV, Ghosh PK (2005) Fouling resistant membranes in desalination and water recovery. Desalination 183(1–3): 301–306

Reuben BG, Perl O, Morgan NL, Stratford P, Dudley LY, Hawes C (1995) Phospholipid coatings for the prevention of membrane fouling. J Chem Technol Biotechnol 63(1):85–91

Revanur R, McCloskey B, Breitenkamp K, Freeman B, Emrick T (2007a) Macromolecules 40:3624

Revanur R, McCloskey B, Breitenkamp K, Freeman BD, Emrick T (2007b) Reactive amphiphilic graft copolymer coatings applied to poly(vinylidene fluoride) ultrafiltration membranes. Macromolecules 40(10):3624–3630. doi:10.1021/ma0701033

Ridgway HF, Rigby MG, Argo DG (1984) Adhesion of a Mycobacterium sp. to cellulose diacetate membranes used in reverse osmosis. Appl Environ Microbiol 47(1):61–67

Rosenberger S, Krüger U, Witzig R, Manz W, Szewzyk U, Kraume M (2002) Performance of a bioreactor with submerged membranes for aerobic treatment of municipal waste water. Water Res 36(2):413–420

Saha NK, Balakrishnan M, Ulbricht M (2009) Fouling control in sugarcane juice ultrafiltration with surface modified polysulfone and polyethersulfone membranes. Desalination 249 (3):1124–1131. doi:10.1016/j.desal.2009.05.013

Saito K, Ito M, Yamagishi H, Furusaki S, Sugo T, Okamoto J (1989) Novel hollow fiber membrane for the removal of metal ion during permeation: preparation by radiation-induced cografting of a crosslinking agent with reactive monomer. Ind Eng Chem Res 28(12):1808–1812

Sata T (1993) Properties of composite membranes formed from ion-exchange membranes and conducting polymers. 4. Change in membrane resistance during electrodialysis in the presence of surface-active agents. J Phys Chem 97(26):6920–6923

Saxena N, Prabhavathy C, De S, DasGupta S (2009) Flux enhancement by argon-oxygen plasma treatment of polyethersulfone membranes. Sep Purif Technol 70(2):160–165

Schulze A, Marquardt B, Kaczmarek S, Schubert R, Prager A, Buchmeiser MR (2010) Electron beam-based functionalization of poly(ethersulfone) membranes. Macromol Rapid Commun 31 (5):467–472

Seman MNA, Khayet M, Hilal N (2012) Comparison of two different UV-grafted nanofiltration membranes prepared for reduction of humic acid fouling using acrylic acid and N-vinylpyrrolidone. Desalination 287:19–29. doi:10.1016/j.desal.2010.10.031

Shah TN, Goodwin JC, Ritchie SMC (2005) Development and characterization of a microfiltration membrane catalyst containing sulfonated polystyrene grafts. J Membr Sci 251(1–2):81–89

Shannon M, Bohn P, Elimelech M, Georgiadis J, Maries B, Mayes A (2008) Science and technology for water purification in the coming decades. Nature 452(7185):301–310

Sheikholaslami R (1999) Composite fouling - inorganic and biological: a review. Environ Prog 18 (2):113–122

Sheikholeslami R (1999) Fouling mitigation in membrane processes. Desalination 123(1):45–53

Shi H, Shi D, Yin L, Luan S, Zhao J, Yin J (2010a) Synthesis of amphiphilic polycyclooctene-graft-poly(ethylene glycol) copolymers by ring-opening metathesis polymerization. React Funct Polym 70(7):449–455

Shi Q, Su Y, Ning X, Chen W, Peng J, Jiang Z (2010b) Graft polymerization of methacrylic acid onto polyethersulfone for potential pH-responsive membrane materials. J Membr Sci 347(1–2):62–68

Shi Q, Su Y, Ning X, Chen W, Peng J, Jiang Z (2011) Trypsin-enabled construction of anti-fouling and self-cleaning polyethersulfone membrane. Bioresour Technol 102(2):647–651. doi:10.1016/j.biortech.2010.08.030

Shi F, Ma Y, Ma J, Wang P, Sun W (2013) Preparation and chacterization of PVDF/TiO$_2$ hybrid membranes with ionic liquid modified nano-TiO$_2$ particles. J Membr Sci 427:259–269

Song YQ, et al (2000) Surface modification of polysulfone membranes by lowtemperature plasma-graft poly(ethylene glycol) onto polysulfone membranes. J Appl Polym Sci 78(5):979–985.

Sotto A, Boromand A, Zhang RX, Luis P, Arsuaga JM, Kim J et al (2011) Effect of nanoparticle aggregation at low concentrations of TiO(2) on the hydrophilicity, morphology, and fouling resistance of PES–TiO(2) membranes. J Colloid Interf Sci 363:540–50

Speaker, LM Chem Abstr (1986) 105, 232980j; U.S. Patent 4,554,076, 1985.

Steen ML, Jordan AC, Fisher ER (2002) Hydrophilic modification of polymeric membranes by low temperature H_2O plasma treatment. J Membr Sci 204(1–2):341–357

Su BH, Fu P, Li Q, Tao Y, Li Z, Zao HS, Zhao CS (2008) Evaluation of polyethersulfone highflux hemodialysis membrane in vitro and in vivo. J Mater Sci—Mater Med 19(2):745–751

Su Q, Pan B, Wan S, Zhang W, Lv L (2010) Use of hydrous manganese dioxide as a potential sorbent for selective removal of lead, cadmium, and zinc ions from water. J Colloid Interface Sci 349(2):607–612

Susanto H, Ulbricht M (2007) Photografted thin polymer hydrogel layers on PES ultrafiltration membranes: characterization, stability, and influence on separation performance. Langmuir 23 (14):7818–7830

Susanto H, Ulbricht M (2008) High-performance thin-layer hydrogel composite membranes for ultrafiltration of natural organic matter. Water Res 42(10–11):2827–2835

Susanto H, Balakrishnan M, Ulbricht M (2007) Via surface functionalization by photograft copolymerization to low-fouling polyethersulfone-based ultrafiltration membranes. J Membr Sci 288(1–2):157–167

Taffarel SR, Rubio J (2010) Removal of Mn2+ from aqueous solution by manganese oxide coated zeolite. Miner Eng 23(14):1131–1138

Taniguchi M, Belfort G (2004) Low protein fouling synthetic membranes by UV-assisted surface grafting modification: Varying monomer type. J Membr Sci 231(1–2):147–157

Taniguchi M, Kilduff JE, Belfort G (2003) Low fouling synthetic membranes by UV-assisted graft polymerization: Monomer selection to mitigate fouling by natural organic matter. J Membr Sci 222(1–2):59–70

Tari SSM, Ali N, Mamat M (2010) Scrutinizing the consequence of surface modification on membrane stability and resistivity towards fouling. In: CSSR 2010–2010 international conference on science and social research, pp741–745

Teng SX, Wang SG, Gong WX, Liu XW, Gao BY (2009) Removal of fluoride by hydrous manganese oxide-coated alumina: Performance and mechanism. J Hazard Mater 168(2–3):1004–1011

Tripathi BP, Dubey NC, Choudhury S, Stamm M (2012) Antifouling and tunable amino functionalized porous membranes for filtration applications. J Mater Chem 22(37):19981–19992

Tripathi BP, Dubey NC, Stamm M (2014) Polyethylene glycol cross-linked sulfonated polyethersulfone based filtration membranes with improved antifouling tendency. J Membr Sci 453:263–274. doi:10.1016/j.memsci.2013.11.007

Turner NW, Jeans CW, Brain KR, Allender CJ, Hlady V, Britt DW (2006) From 3D to 2D: a review of the molecular imprinting of proteins. Biotechnol Prog 22(6):1474–1489

Tyszler D, Zytner RG, Batsch A, Brügger A, Geissler S, Zhou H, Klee D, Melin T (2006) Reduced fouling tendencies of ultrafiltration membranes in wastewater treatment by plasma modification. Desalination 189(1-3 SPEC. ISS.):119–129

Ulbricht M, Belfort G (1996) Surface modification of ultrafiltration membranes by low temperature plasma II. Graft polymerization onto polyacrylonitrile and polysulfone. J Membr Sci 111(2):193–215

Ulbricht M, Hicke H (1993) Photomodification of ultrafiltration membranes, 1. Photochemical modification of polyacrylonitirle ultrafiltration membranes with aryl azides. Angew Makromol Chem 210:69–95

Ulbricht M, Riedel M, Marx U (1996) Novel photochemical surface functionalization of polysulfone ultrafiltration membranes for covalent immobilization of biomolecules. J Membr Sci 120(2):239–259

Ulbricht M, Riedel M (1998) Ultrafiltration membrane surfaces with grafted polymer 'tentacles': Preparation, characterization and application for covalent protein binding. Biomaterials 19 (14):1229–1237

Van Der Bruggen B (2009a) Chemical modification of polyethersulfone nanofiltration membranes: a review. J Appl Polym Sci 114(1):630–642

Van Der Bruggen B (2009b) Comparison of redox initiated graft polymerisation and sulfonation for hydrophilisation of polyethersulfone nanofiltration membranes. Eur Polymer J 45(7):1873–1882. doi:10.1016/j.eurpolymj.2009.04.017

Venault A, Liu YH, Wu JR, Yang HS, Chang Y, Lai JY, Aimar P (2014) Low-biofouling membranes prepared by liquid-induced phase separation of the PVDF/polystyrene-b-poly (ethylene glycol) methacrylate blend. J Membr Sci 450:340–350

Wang Y, Su Y, Sun Q, Ma X, Ma X, Jiang Z (2006a) Improved permeation performance of Pluronic F127-polyethersulfone blend ultrafiltration membranes. J Membr Sci 282(1–2):44–51

Wang YQ, Su YL, Ma XL, Sun Q, Jiang ZY (2006b) Pluronic polymers and polyethersulfone blend membranes with improved fouling-resistant ability and ultrafiltration performance. J Membr Sci 283(1–2):440–447

Wang YQ, Wang T, Su YL, Peng FB, Wu H, Jiang ZY (2006c) Protein-adsorption-resistance and permeation property of polyethersulfone and soybean phosphatidylcholine blend ultrafiltration membranes. J Membr Sci 270(1–2):108–114

Wang H, Yang L, Zhao X, Yu T, Du Q (2009a) Improvement of hydrophilicity and blood compatibility on polyethersulfone membrane by blending sulfonated polyethersulfone. Chin J Chem Eng 17(2):324–329. doi:10.1016/S1004-9541(08)60211-6

Wang Z, Wu Z, Tang S (2009b) Extracellular polymeric substances (EPS) properties and their effects on membrane fouling in a submerged membrane bioreactor. Water Res 43(9):2504–2512

Wang D, Zou W, Li L, Wei Q, Sun S, Zhao C (2011) Preparation and characterization of functional carboxylic polyethersulfone membrane. J Membr Sci 374(1–2):93–101. doi:10.1016/j.memsci.2011.03.021

Wang Z, Wang H, Liu J, Zhang Y (2014) Preparation and antifouling property of polyethersulfone ultrafiltration hybrid membrane containing halloysite nanotubes grafted with MPC via RATRP method. Desalination 344:313–320

Wavhal DS, Fisher ER (2002a) Hydrophilic modification of polyethersulfone membranes by low temperature plasma-induced graft polymerization. J Membr Sci 209(1):255–269

Wavhal DS, Fisher ER (2002b) Modification of porous poly(ether sulfone) membranes by low-temperature CO_2-plasma treatment. J Polym Sci Part B: Polym Phys 40(21):2473–2488

Wavhal DS, Fisher ER (2003) Membrane surface modification by plasma-induced polymerization of acrylamide for improved surface properties and reduced protein fouling. Langmuir 19 (1):79–85

Wei J, Helm GS, Corner-Walker N, Hou X (2006) Desalination, 192, 252

Wei S, Jakusch M, Mizaikoff B (2006a) Capturing molecules with template materials—analysis and rational design of molecularly imprinted polymers. Anal Chim Acta 578:50–58

Wei X, Wang R, Li Z, Fane AG (2006b) Development of a novel electrophoresis-UV grafting technique to modify PES UF membranes used for NOM removal. J Membr Sci 273(1–2):47–57. doi:10.1016/j.memsci.2005.11.049

Whitcombe M, Vulfson E (2001) Imprinted polymers. Adv Mater 13:467–478

Wu J, Huang X (2008) Effect of dosing polymeric ferric sulfate on fouling characteristics, mixed liquor properties and performance in a long-term running membrane bioreactor. Sep Purif Technol 63(1):45–52

Wu G, Gan S, Cui L, Xu Y (2008) Preparation and characterization of PES/TiO_2 composite membranes. Appl Surf Sci 254(21):7080–7086

Wulff G (1995) Molecular imprinting in crosslinked materials with the aid of molecular templates—a way towards aritificial antibodies. Angew chemInt ED Engl 34:1812–1832

Wulff G (2002) Enzyme-like catalysis by molecularly imprinted polymers. Chem Rev 102(1):1–27

Xi W, Rong W, Zhansheng L, Fane A (2006) Development of a novel electrophoresis-UV grafting technique to modify PES UF membranes for NOm removal. Membr Sci 273:47–57

Xiang T, Wang LR, Ma L, Han ZY, Wang R, Cheng C, Xia Y, Qin H, Zhao CS (2014) From commodity polymers to functional polymers. Scientific Reports 3. doi:10.1038/srep04604

Xie Y, Li SS, Jiang X, Xiang T, Wang R, Zhao CS (2015) Zwitterionic glycosyl modified polyethersulfone membranes with enhanced anti-fouling property and blood compatibility. J Colloid Interface Sci 443:36–44. doi:10.1016/j.jcis.2014.11.053

Xu FJ, Kang ET, Neoh KG (2005) UV-induced coupling of 4-vinylbenzyl chloride on hydrogen-terminated Si(100) surfaces for the preparation of well-defined polymer-Si hybrids via surface-initiated ATRP. Macromolecules 38(5):1573–1580. doi:10.1021/ma049225a

Yamagishi H, Crivello JV, Belfort G (1995a) Development of a novel photochemical technique for modifying poly(arylsulfone) ultrafiltration membranes. J Membr Sci 105(3):237–247

Yamagishi H, Crivello JV, Belfort G (1995b) Evaluation of photochemically modified poly (arylsulfone) ultrafiltration membranes. J Membr Sci 105(3):249–259

Yan S, Maeda H, Kusakabe K, Morooka S (1994) Thin palladium membrane formed in support pores by metal-organic chemical vapor deposition method and application to hydrogen separation. Ind Eng Chem Res 33(3):616–622

Yan MG, Liu LQ, Tang ZQ, Huang L, Li W, Zhou J, Gu JS, Wei XW, Yu HY (2008) Plasma surface modification of polypropylene microfiltration membranes and fouling by BSA dispersion. Chem Eng J 145(2):218–224

Yang B, Yang W (2003) Photografting modification of PET nuclepore membranes. Macromol Sci A Pure Appl Chem 40:309–320

Yang W, Cicek N, Ilg J (2006) State-of-the-art of membrane bioreactors: worldwide research and commercial applications in North America. J Membr Sci 270(1–2):201–211

Yang YF, Wan LS, Xu ZK (2009a) Surface hydrophilization for polypropylene microporous membranes: a facile interfacial crosslinking approach. J Membr Sci 326(2):372–381. doi:10. 1016/j.memsci.2008.10.011

Yang YF, Wan LS, Xu ZK (2009b) Surface hydrophilization of microporous polypropylene membrane by the interfacial crosslinking of polyethylenimine. J Membr Sci 337(1–2):70–80. doi:10.1016/j.memsci.2009.03.023

Ye L, Mosbach K (2008) Molecular imprinting: synthetic materials as substitutes for biological antibodies and receptors. Chem Mater 20(3):859–868

Yin J, Coutris N, Huang Y (2012) Experimental investigation of aligned groove formation on the inner surface of polyacrylonitrile hollow fiber membrane. J Membr Sci 394–395:57–68

Ying L, Kang ET, Neoh KG (2002) Covalent immobilization of glucose oxidase on microporous membranes prepared from poly(vinylidene fluoride) with grafted poly(acrylic acid) side chains. J Membr Sci 208(1–2):361–374

Yu HY, Hu MX, Xu ZK, Wang JL, Wang SY (2005) Surface modification of polypropylene microporous membranes to improve their antifouling property in MBR: NH3 plasma treatment. Sep Purif Technol 45(1):8–15

Yu HY, Xu ZK, Xie YJ, Liu ZM, Wang SY (2006) Flux enhancement for polypropylene microporous membrane in a SMBR by the immobilization of poly(N-vinyl-2-pyrrolidone) on the membrane surface. J Membr Sci 279(1–2):148–155

Yu HY, He XC, Liu LQ, Gu JS, Wei XW (2007) Surface modification of polypropylene microporous membrane to improve its antifouling characteristics in an SMBR: N2 plasma treatment. Water Res 41(20):4703–4709

Yu HY, Liu LQ, Tang ZQ, Yan MG, Gu JS, Wei XW (2008a) Mitigated membrane fouling in an SMBR by surface modification. J Membr Sci 310(1–2):409–417

Yu HY, Liu LQ, Tang ZQ, Yan MG, Gu JS, Wei XW (2008b) Surface modification of polypropylene microporous membrane to improve its antifouling characteristics in an SMBR: air plasma treatment. J Membr Sci 311(1–2):216–224

Yu H, Cao Y, Kang G, Liu J, Li M, Yuan Q (2009a) Enhancing antifouling property of polysulfone ultrafiltration membrane by grafting zwitterionic copolymer via UV-initiated polymerization. J Membr Sci 342(1–2):6–13

Yu LY, Xu ZL, Shen HM, Yang H (2009b) Preparation and characterization of PVDF-SiO2 composite hollow fiber UF membrane by sol-gel method. J Membr Sci 337(1–2):257–265

Yu H, et al (2013) Development of a hydrophilic PES ultrafiltration membrane containing SiO2@N-Halamine nanoparticles with both organic antifouling and antibacterial properties. Desalination. 326:69–76.

Yun S, Ted Oyama S (2011) Correlations in palladium membranes for hydrogen separation: a review. J Membr Sci 375(1–2):28–45

Yune PS, Kilduff JE, Belfort G (2011) Fouling-resistant properties of a surface-modified poly (ether sulfone) ultrafiltration membrane grafted with poly(ethylene glycol)-amide binary monomers. J Membr Sci 377(1–2):159–166. doi:10.1016/j.memsci.2011.04.029

Zhan J, Liu Z, Wang B, Ding F (2004) Modification of a membrane surface charge by a low temperature plasma induced grafting reaction and its application to reduce membrane fouling. Sep Sci Technol 39(13):2977–2995

Zhang H, Ye L, Mosbach K (2006) Non-covalent molecular imprinting with emphasis on its application in separation and drug development. Mol Recognit 19:248–259

Zhang M, Nguyen QT, Ping Z (2009) Hydrophilic modification of poly (vinylidene fluoride) microporous membrane. J Membr Sci 327(1–2):78–86

Zhang J, Zhang M, Zhang K (2014) Fabrication of poly(ether sulfone)/poly(zinc acrylate) ultrafiltration membrane with anti-biofouling properties. J Membr Sci 460:18–24

Zhao B, Brittain WJ (2000) Polymer brushes: surface-immobilized macromolecules. Prog Polym Sci (Oxf) 25(5):677–710

Zhao C, Liu X, Rikimaru S, Nomizu M, Nishi N (2003) Surface characterization of polysulfone membranes modified by DNA immobilization. J Membr Sci 214(2):179–189

Zhao ZP, Li J, Wang D, Chen CX (2005) Nanofiltration membrane prepared from polyacrylonitrile ultrafiltration membrane by low-temperature plasma: 4. Grafting of N-vinylpyrrolidone in aqueous solution. Desalination 184(1–3):37–44

Zhao K, Cheng G, Huang J, Ying X (2008) Rebinding and recognition properties of protein-macromolecularly imprinted calcium phosphate/alginate hybrid polymer microspheres. React Funct Polym 68(3):732–741

Zhao W, Huang J, Fang B, Nie S, Yi N, Su B, Li H, Zhao C (2011) Modification of polyethersulfone membrane by blending semi-interpenetrating network polymeric nanoparticles. J Membr Sci 369(1–2):258–266

Zhao C, Xue J, Ran F, Sun S (2013a) Modification of polyethersulfone membranes—a review of methods. Prog Mater Sci 58(1):76–150

Zhao W, Mou Q, Zhang X, Shi J, Sun S, Zhao C (2013b) Preparation and characterization of sulfonated polyethersulfone membranes by a facile approach. Eur Polymer J 49(3):738–751

Zhao S, Yan W, Shi M, Wang Z, Wang J, Wang S (2015) Improving permeability and antifouling performance of polyethersulfone ultrafiltration membrane by incorporation of ZnO-DMF dispersion containing nano-ZnO and polyvinylpyrrolidone. J Membr Sci 478:105–116. doi:10.1016/j.memsci.2014.12.050

Zhou R, Ren PF, Yang HC, Xu ZK (2014) Fabrication of antifouling membrane surface by poly (sulfobetaine methacrylate)/polydopamine co-deposition. J Membr Sci 466:18–25

Zhu LP, Xu L, Zhu BK, Feng YX, Xu YY (2007a) Preparation and characterization of improved fouling-resistant PPESK ultrafiltration membranes with amphiphilic PPESK-graft-PEG copolymers as additives. J Membr Sci 294(1–2):196–206

Zhu LP, Zhang XX, Xu L, Du CH, Zhu BK, Xu YY (2007b) Improved protein-adsorption resistance of polyethersulfone membranes via surface segregation of ultrahigh molecular weight poly(styrene-alt-maleic anhydride). Colloids Surf B 57(2):189–197

Zhu L, Wang J, Zhu B, Xu Y (2008a) Molecular design and synthesis of amphiphilic copolymers, and the performances of their blend membranes. Acta Polymerica Sinica 4:309–317

Zhu LP, Yi Z, Liu F, Wei XZ, Zhu BK, Xu YY (2008b) Amphiphilic graft copolymers based on ultrahigh molecular weight poly(styrene-alt-maleic anhydride) with poly(ethylene glycol) side chains for surface modification of polyethersulfone membranes. Eur Polymer J 44(6):1907–1914

Zhu J, Guo N, Zhang Y, Yu L, Liu J (2014) Preparation and characterization of negatively charged PES nanofiltration membrane by blending with halloysite nanotubes grafted with poly (sodium 4-styrenesulfonate) via surface-initiated ATRP. J Membr Sci 465:91–99

Zinadini S, Zinatizadeh AA, Rahimi M, Vatanpour V, Zangeneh H (2014) Preparation of a novel antifouling mixed matrix PES membrane by embedding graphene oxide nanoplates. J Membr Sci 453:292–301

Chapter 5
Membrane Characterization Techniques

Membrane characterization is critical at various stages in the lifecycle of a membrane. At the research and development stage, it forms a critical element of the iterative design-synthesise-test-evaluate process, while in operating the characterization of membranes is more limited and usually relates to determining whether cleaning/regeneration is required, or eventually membrane replacement. Membrane characterization is critical at various stages in the lifecycle of a membrane. At the research and development stage, it forms a critical element of the iterative design-synthesise-test-evaluate process, while in operating the characterization of membranes is more limited and usually relates to determining whether cleaning/regernation is required, or eventually membrane replacement. Finally, membranes are also characterised in postmortems, which give critical insight into membrane failure and help propose modifications to membrane preparations or operation. This chapter provides a critical review of membrane characterization, particularly PES characterization and some related comments on membrane stability.

5.1 Introduction

After introducing new components into the membrane matrix using different chemical or physical strategies, the structure and morphology of the membranes may relatively change. Therefore, after surface modification process of membranes, characterisation of membranes is a paramount step to confirm if the alterations of membrane structure, and membrane morphology and membrane performance are applicable. There are three types of characterization of membranes:

© Springer Nature Singapore Pte Ltd. 2017
B. Ladewig and M.N.Z. Al-Shaeli, *Fundamentals of Membrane Bioreactors*,
Springer Transactions in Civil and Environmental Engineering,
DOI 10.1007/978-981-10-2014-8_5

(1) characterisation of composition, (2) characterisation of morphology and structure, and (3) characterisation of performance (Zhao et al. 2013).

5.2   Composition Characterization

After modification the membrane surface and introducing new components into its matrix, the bulks or compositions of membrane surfaces membrane may change to some extents. This change can be detected by different means as described below.

5.3   Fourier Transform Infrared Spectroscopy (FTIR)

Fourier transform infrared spectroscopy (FTIR) is an effective tool used to identify material chemistry and produce an infrared spectrum of absorption, photoconductivity, mission or Raman scattering of a solid, liquid or gas. A FTIR spectrometer detects high spectral resolution data over a wide spectral range. This is a significant advantage over dispersive spectrometer, which measures intensity over a narrow range of wavelengths at a time. FTIR deals with the vibration of various types of molecular bonds at different frequencies. When exposing material into infrared radiation, the molecules would absorb some of the infrared energy. This could happen when the radiation frequency provides energy in an accurate amount, which is required by one of the bonds in the molecules. In Membranes, many research studies have been conducted to characterise the chemical structure of the modified PES membranes using ATR-FTIR as analytical tool (Pieracci et al. 1999; Fontyn et al. 1986b; Millesime et al. 1994). For example, (Fontyn et al. 1986a) investigated the chemical characterisation of PES and PVDF ultrafiltration membranes using spectroscopic strategies. They state that FTIR can be utilised as a tool to evaluate PES/PVDF membrane component after introducing new additives into their membrane matrix. In a research study conducted by (Millesime et al. 1994), PES membranes were coated with polyvinyl imidazole (PVI) and then quaternized using a cross-linking agent bearing Bisphenol A group. The thickness of PVI was calculated using ATR-FTIR spectroscopy. (Pieracci et al. 1999) modified commercial PES membranes (10 KDa) by photografting of three hydrophilic monomers (*N*-vinyl-2-pyrrolidinone (NVP), *N*-vinylcaprolactam (NVC), and *N*-vinylformamide (NVF). They used ATR-FTIR in their study to confirm the occurrence of photochemical grafting of the hydrophilic monomers onto the surface of PES UF membranes as shown in Figs. 5.1 and 5.2.

As shown in Fig. 5.1, Compared with neat PES membranes (10 KDa), there is a wide peak between 3200–3600 cm^{-1}, ascribed to the stretch of OH group in all the modified membranes. (Pieracci et al. 1999) also state that after irradiation process, the amount of hydroxyl groups formed on the surface of membrane has been linearly increased with irradiation time (1, 3, 5, 7 and 10 min).

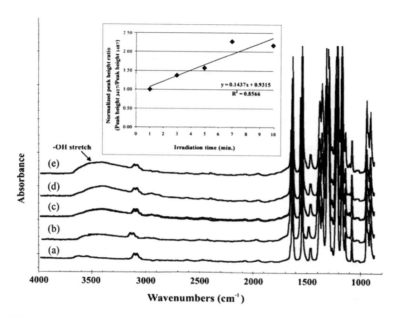

Fig. 5.1 FTIR/ATR spectra with a germanium crystal (45°) of the surface of **a** Pristine PES, **b** a PES membrane irradiated for 1 min, **c** a PES membrane irradiated for 3 and 5 min **d** a PES membrane irradiated for 7 min **e** a PES membrane irradiated for 10 min (Pieracci et al. 1999)

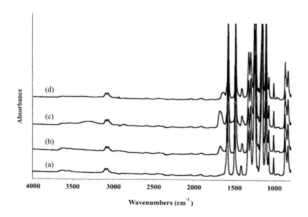

Fig. 5.2 FTIR/ATR spectra with a germanium crystal (45°) of the surface of PES: **a** unmodified, **b** PES membrane modified with 1 wt% N-vinyl-2-pyrrolidinone (NVP) and irradiated for 3 min, **c** PES membrane modified with 0.75 wt% N-vinylformamide (NVF) and 10 min of irradiation, and **d** PES membrane modified with 0.1 wt% N-vinylcaprolactam (NVC) and 5 min of irradiation (Pieracci et al. 1999)

As shown in Fig. 5.2, compared to the pristine PES, there is an absorbance band, appearing at approximately 1678 cm^{-1}, assigned to the amide I carbonyl stretch of the NVP five-membered lactam ring and NVC seven-membered ring. The amide I carbonyl band of NVF was shifted down to ~1645 cm^{-1} because the carbonyl group was not within a ring structure. The occurrence of amide I carbonyl stretch bands in all three spectra confirmed that the monomers had been successfully grafted onto the surface of the PES membrane. To sum up, FTIR-ATR is not very surface sensitive due to the large penetration depth, although it is vey common technique used to characterise or analyse the membrane surface quickly.

5.4 Nuclear Magnetic Resonance (NMR)

NMR is one of the principal techniques used to achieve detailed information about the chemical structure, topology of molecules. In membranes, it is rarely used to directly characterise the chemical structure of the surface-modified membranes. However, it is usually used to characterise the structure of the modifying agent (Zhao et al. 2011a; Brayfield et al. 2008; Yi et al. 2010b; Gaina et al. 2011). (Yi et al. 2010b) have been modified PES membranes with amphiphilic polysulfone-*graft*-poly (ethylene glycol) methyl ether methacrylate (PSF-*g*-POEM). NMR was used to characterise the amphiphilic additive (PSF-g-POEM).

5.5 X-ray Photoelectron Spectroscopy (XPS)

X-ray photoelectron spectroscopy (XPS) (known as electron microscopy) is a highly quantitative spectroscopic technique used to characterise the chemical distribution and structure composition of the uppermost atomic surface of the samples (Hester et al. 1999; Liu and Kim 2011). It can be also used to recognise the elemental composition, chemical state (which is very important for the main elements of polymer such as C, O, N, S and F) and electronic state of the elements. In membranes, it is also used for characterising the modified PES membranes (Mosqueda-Jimenez et al. 2006; Chu et al. 2005; Liu et al. 2009a; Liu and Kim 2011; Wu and Yang 2006; Wei et al. 2012; Li et al. 2005).

Lin et al. (2015) synthesised tris(2,4,6-trimethoxyphenyl) poly(phenylene oxide)-methylene quaternary-phosphonium-bromide (TPPOQP-Br) and triethyl poly(phenylene oxide)-methylene quaternary-phosphonuim-bromide (TPPOQA-Br). XPS test have been conducted to verify the effects of TPPOQP-Br and TPPOQA-Br on BPPO UF membranes. The results indicated that the composite membranes (BPPO/TPPOQP-Br) showed an increase in the concentration of TPPOQP-Br from the top surface to the bottom surface. Conversely, an opposite concentration gradient of TPPOQA-Br was observed for BPPO/TPPOQA-Br composite membranes.

5.6 Thermal Gravitational Analysis (TGA)

Generally, thermal gravitational analysis is an analytical technique, which is used to measure weight changes in a material as a function of temperature (with constant heating rate) or time (with constant mass loss or constant temperature) under a controlled atmosphere. It is mainly used to evaluate the thermal stability of material and its composition. In membrane field, TGA has been commonly used to evaluate the thermal stability of membranes or thermal degradation behaviour of membranes. Zhang et al. (2014) modified PES UF membranes with Poly (zinc acrylate) (PZA) to tune the hydrophilicity and improve the anti-biofouling properties of membranes. They used TGA to prove the presence PZA within the membrane matrix.

5.7 Differential Scanning Calorimetry (DSC)

DSC is an analytical technique, which is used thoroughly to measure the difference in the amount of heat required to increase the temperature of a sample and reference as a function of temperature. In membrane fields, many researchers have used DSC for different purposes, for example, (Bolong et al. 2009; Yi et al. 2010b; Li et al. 2005) used DSC in their study to find the miscibility between PES and the additives via analysing the glass transition temperature of the polymers. For sample preparation, membrane samples were cut into shape; accuracy amount of membrane sample was measured in digital balance (weight should less than 10 mg). The membrane sample was then folded and placed into preweighed aluminium crucible (pan) and after that kept sample as thin as possible and covered as much as the pan bottom as possible. Then, place the lid on the pan and both the lid and the pan (with sample inside) assembly were placed in the well of the lower crimping die. Thereafter firmly press down on the plunger and then remove the pan with tweezers and make sure that the sample has been completely encapsulated in the pan. The sample was measured under nitrogen atmosphere.

To sum up, FTIR, NMR, XPS, TGA and DSC are very important tools used to indicate the alteration on the membrane surface after surface modification using chemical or physical methods.

However, introducing new components into PES matrix to modify its surface may not only result in chemical or physical alterations, but also affect the membrane morphology and structure of membrane. Therefore, characterization the structure and morphology of PES membrane after surface modification is very important.

5.8 Energy-Dispersive X-ray Spectroscopy (EDS)

Energy-dispersive-ray spectroscopy (EDS) elemental microanalysis in conjunction with scanning electron microscope (SEM) has been extensively used for characterising the distribution of elements on the surface of samples. It has a major advantage in that the spatial resolution can be quite high and SEM images are obtained for the same area being analysed, so that information regarding both the surface features/feature size can be combined with knowledge of the elemental composition.

Characterisation of morphology and structure

Direct visualisation of structural features with the aid of microtechnique has the advantage of being a model-free method, which generate abundant, practical and direct information about the membrane. Microtechnique has been commonly used to characterise the structure and morphology of the untreated and treated membranes. Microtechniques include Scanning Electron Microscopy (SEM), Atomic Force Microscopy (AFM) and so on.

5.9 Scanning Electron Microscopy (SEM)

SEM has been considered as a powerful characterisation tool used commonly by many researchers to characterise membrane properties quantitatively (i.e. porosity, pore size, pore shape, and pore density) and qualitatively (i.e. visual or direct observation). It has been practically used for 30 years and many innovations and advancement are continuously added to it. Many journal articles have been published on using SEM to characterise membrane surface properties (PES or PVDF membranes) including porosity, pore size, pore size distribution, pore density, pore shape, surface roughness and fouling (Liu et al. 2009b; Mu and Zhao 2009; Mansourpanah et al. 2009a, b, 2010; Mosqueda-Jimenez et al. 2004; Zhao et al. 2011a, b; Razmjou et al. 2011; Kochan et al. 2010; Dejeu et al. 2009; Rahimpour 2011; Rahimpour et al. 2008; Yi et al. 2010a, b; Wei et al. 2012; Li et al. 2005; Peeva et al. 2012). SEM has advantages and limitations. The advantages are: 1) characterise the membrane by providing practical and direct structural information about the membrane, leading to impressive analytical output. Thus, SEM has become a powerful technique of characterisation not only for research but also for quality and product. The limitations of SEM are: (1) SEM is extremely restricted to examine very small area (2) SEM requires the membrane samples to be dried before testing. In some samples, drying generates undesirable structure collapse. (Mu and Zhao 2009) used thermal-induced surface cross-linking method to modify PES porous membranes. They used poly (ethylene glycol) diacrylate (PEGDA) (hydrophilic additive) as hydrophilic modifier. Characterising the morphology of PES membranes is conducted by SEM as present schematically in Fig. 5.3.

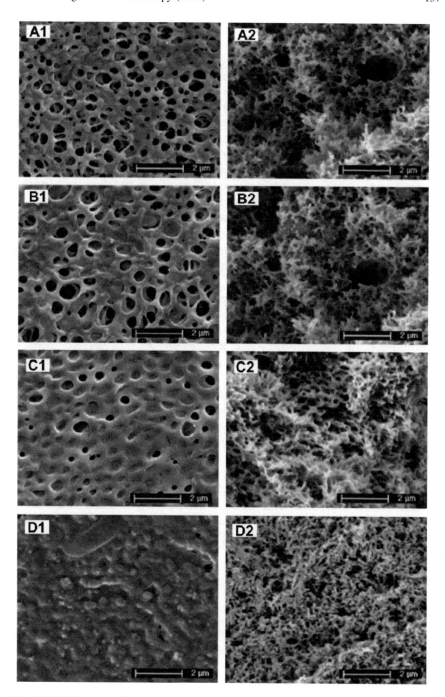

Fig. 5.3 SEM images of the pristine and modified PES membranes: **a** pristine PES membrane; **b–d** the modified PES membranes with the mass gain of 115.4, 253.1, 345.6 $\mu g/cm^2$, respectively. (*1*) denotes the separation surface, (*2*) is the cross-section (Mu and Zhao 2009)

As can be seen from Fig. 5.3, pristine PES membrane (A1) showed a micro-filtration porous surface. The pore size of membrane surface is (0.2–0.4 μm) and the cross-section (A2) of PES membrane typically showed network structure. After membrane modification, the cross-linker modifier PEGDA decreases the pore size on surface of membrane. Increasing mass gain (115.4, 253.1, 345.6 μg/cm^2) leads to the gradual decrease of both pore size and number because of the coverage of cross-linking layer. No pore is approximately noted on the surface of the membrane when the mass gain value is 345.6 μg/cm^2. No pronounced difference is found in the cross-section morphologies between neat PES membranes (A1) and the treated PES membranes (B2, C2) with lower mass gains. However, with higher mass gain, the cross-section morphology of the treated PES membrane (D2) looks to be denser, indicating that the modification may reach membrane bulk.

In another research study, (Rahimpour et al. 2008) modified PES with TiO$_2$ nanoparticles using UV irradiation method. Scanning electron microscopy was conducted to characterise membrane properties (i.e. pore size and porosity). They concluded that porosity and pore size of the treated PES membrane was higher than the virgin PES membrane.

Field emission scanning electron microscopy (FESEM) is another analytical technique used to investigate the structures of the molecular surface. In membranes, due to higher resolution than other microscopic techniques (e.g. SEM), it is also used to inspect membrane surface (e.g. surface porosity, pore size and its distri-bution) and cross-sectional morphologies (Bolong et al. 2009). The cross-sectional regions of membranes can be prepared by fracturing the membranes in liquid nitrogen. Then, samples were mounted either onto the plate of the specimen sample holder (stub) (for surface observation) or side stage of the specimen sample holder (for cross-section observation) using conductive double side tape, then sputter coated the samples with different coating metals (platinum, gold, iridium and gold-palladium). The coated samples were scrutinised using FESEM with different resolution and different magnification.

Conversely, Transmission Electron Microscopy (TEM) is another technique used to inspect finer detail even as small than a single column of atoms. In membrane, it is seldom utilised to characterise PES membranes after modifications. The advantage of TEM is its outstandingly higher resolution (less than 1 nm level).

5.10   Atomic Force Microscopy

Atomic force microscopy (AFM) is a versatile tool used to characterise the surfaces physically. It is a form of scanning probe microscopy (SPM) with a very high resolution; up to the order of fractions of a nanometer and it is more than 1000 times better than the optical diffraction limit. The AFM is considered as one of the foremost tools for imaging, measuring, and manipulating matter at the nano-scale. In the membrane, AFM has been used to characterise ultrafiltration membranes and to a lesser degree characterise microfiltration membranes. AFM is also used to measure the pore size, porosity, pore size distribution, aggregate size and nodule size at the membrane surface. In recent years, AFM was also used to study the adhesion properties of membrane surfaces using force measurement (force mode). And colloidal probe technique to quantify the interaction force acting between

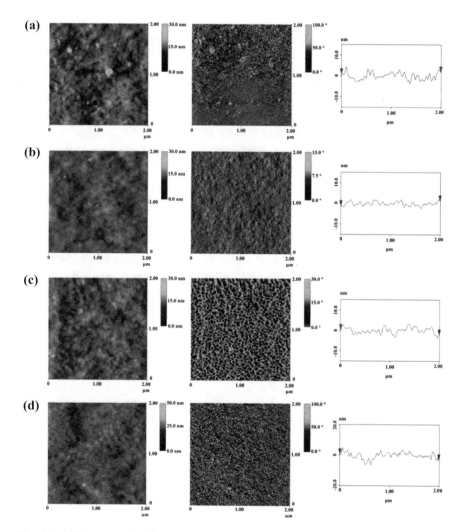

Fig. 5.4 AFM topography, phase image, and cross-sectional roughness of thin-film composites by interfacial polymerization: **a** virgin PES membrane, **b** film composite with PVA, **c** film composite with PEG, and **d** film composite with chitosan. The root mean square roughness is (**a**) 2.1 nm, (**b**) 1.2 nm, (**c**) 1.5 nm, and (**d**) 2.0 nm. Image size was 2×2 μm (Liu et al. 2009a)

surface and probe is used, in which micrometre-sized spheres are attached to the AFM cantilever. Hydrophilic membranes are strongly interacting with hydrophilic probe as indicated by a large phase shift, whereas the hydrophobic membranes give only a small phase shift (Rana and Matsuura 2010).

There are three modes to operate AFM: (1) contact mode (static mode) (2) tapping mode (3) noncontact mode. Tapping and noncontact mode (which are also called dynamic modes) are usually used to characterise the surface topography and phase image of PES membranes (Mansourpanah et al. 2009a, 2010; Razmjou et al. 2011; Liu et al. 2009a; Abu Seman et al. 2010; Rahimpour 2011; Wu and Yang 2006; Tur et al. 2012).

Liu et al. (2009a) grafted PES membranes by three hydrophilic polymers (Polyvinyl alcohol (PVA), Polyethylene glycol (PEG) and chitosan) through interfacial grafting technique; then they characterise the virgin and modified PES membranes by AFM (see Fig. 5.4). As presented in Fig. 5.4, pristine PES membranes clearly exhibited ridge-and-valley structure. The bar indicated the vertical deviations in the surface of membrane, so the white area represents the highest level, whilst the dark region represents the lowest level. PES membrane surface seemed to be heterogeneous (not homogenous), while the surface-modified PES membranes showed more or less surface roughness than the pristine PES membrane.

Low et al. (2014) synthesised two-dimensional zeolitic imidazolate framework with leaf-shaped morphology (ZIF-L) and blend it with PES UF membranes. AFM test was conducted to verify the effects of ZIF-L on the roughness of modified membranes. The results showed that the roughness values in the projected area of 1×1 μm decreased from 5.99 to 3.55 nm as 0.5 % of ZIF-L was added to the casting solution. It was well documented that lower surface roughness lead to higher antifouling properties, as foulants were likely to be absorbed in the valleys of the membranes with rough surface (Vrijenhoek et al. 2001).

5.11 Characterization of Membrane Performance

Characterization of membrane performance after surface modification is paramount. The characteristics of modified membranes include hydrophilicity/hydrophobicity, permeability (flux) and selectivity, antifouling properties and so forth.

5.12 Hydrophilicity/Hydrophobicity of Membranes

Hydrophilicity is one of the most important aspects to characterise membrane performance. Water contact angle measurement is widely used to assess the hydrophilicity and the wetting properties of the membrane surface. However, the measurement value cannot accurately act as the indicator comparing different membranes as this value does not solely rely on surface hydrophilicity but also porosity, surface roughness, pore size and pore size distribution of the membranes (Rana and Matsuura 2010). Nevertheless, the static contact angle changes with the age of the water drop which can be a better index for surface hydrophilicity (Zhao et al. 2008c). Clearly, the higher hydrophilic porous membrane not only has a smaller water contact angle but also a rapid decrease rate of static contact angle. The techniques that are used to measure contact angle include statistic sessile drop method (Mansourpanah et al. 2009a, 2010; Zhao et al. 2011a; Cao et al. 2010; Bolong et al. 2009; Liu et al. 2009a; Pieracci et al. 1999; Rahimpour et al. 2008; Yi et al. 2010a; Wei et al. 2012; Pazokian et al. 2011) and captive air bubble technique (Pieracci et al. 1999; Tur et al. 2012). To ensure the accuracy, several determinations at different locations were tested for each membrane sample and an average was taken.

Many researchers have been used as contact angle measurement in their study to investigate the hydrophilicity of PES membranes (Mu and Zhao 2009; Mansourpanah et al. 2009a, b, 2010; Mosqueda-Jimenez et al. 2006; Van der Bruggen 2009; Zhao et al. 2011a, b; Bolong et al. 2009; Chu et al. 2005; Liu et al. 2009a; Pieracci et al. 1999; Rahimpour et al. 2008; Gaina et al. 2011; Liu and Kim 2011; Wu and Yang 2006; Wei et al. 2012; Yi et al. 2010a; Tur et al. 2012; Pazokian et al. 2011). They concluded that introduction of hydrophilic components into PES membrane could decrease the water contact angles. However, many researchers noted that the contact angles kept changing after the water was dropped on the PES membrane surface. (Bolong et al. 2009) attribute it to evaporation effect and recommended that the measurements should be made as faster as possible (less than 10 s) to reduce evaporation effect. (Mu and Zhao 2009) also observed that the water contact angle of the virgin PES membrane was significantly declined from about 89 °C at the drop time of 0th second to 66 °C at 185th seconds. They ascribed this phenomenon to the permeation of water drop into membrane pore.

5.13 Permeability and Selectivity of Membranes

The permeability of membrane is very crucial step after modification of membranes, which many researchers are preferred. The permeation of modified membranes is often verified by measurement of water (flux) or measurement of gas flux (Susanto and Ulbricht 2007; Mansourpanah et al. 2009a, b, 2010; Mosqueda-Jimenez et al. 2006; Van der Bruggen 2009; Zhao et al. 2011a, b; Chu et al. 2005; Pieracci et al. 1999; Saha et al. 2009; Rahimpour 2011; Rahimpour et al. 2008; Wei et al. 2012; Peeva et al. 2012; Tur et al. 2012). The alteration and recovery of water permeability for stimulus-response membranes were also conducted under various environmental stimuli (e.g. ionic strength, pH, temperature).

To obtain steady flux, the membrane must be pre-compacted first by DDI water for at least 1 h before calculation flux to completely make sure the water is immersed into the membranes (Zhao et al. 2011b). Constant transmembrane pressure mode is usually used to determine the membrane permeation.

Polyethylene glycols (PEGs) with different molecular weights (e.g. 20,35,100,200,300,400 KDa and so on), poly (ethylene oxide) with different molecular weights or bovine serum albumin (BSA) are commonly three-marker macromolecule used to characterise the selectivity of the polymeric membranes (Mu and Zhao 2009; Pieracci et al. 1999; Yi et al. 2010b; Huang et al. 2011). They also used to determine the solute rejection and molecular weight cut-off (MWCO). MWCO of membranes is the lowest molecular weight of the marker macromolecule (e.g. PEG) at a rejection of 90 % in the measurements. Pre-size of membranes is determined based on the MWCO using the following equation.

$$r = 0.026 \, \text{nm} \sqrt{\text{MWCO}(\frac{g}{\text{gmol}})} - 0.03 \, \text{nm}$$

5.14 Antifouling Properties

Membrane fouling is a repugnant problem in all membrane processes, particularly membrane bioreactors. Therefore, the modification of membranes is very important step toward increasing the antifouling properties of membranes. Moreover, characterisation of antifouling properties after surface modification is very necessary. Flux and BSA (foulant) decline highlights the severity of antifouling. Thus, flux and protein are solute used as index for antifouling property (Mu and Zhao 2009; Pieracci et al. 1999; Yi et al. 2010b; Huang et al. 2011). Therefore, antifouling properties of membranes are related to the nature of material, surface hydrophilicity, surface charge and the roughness of membrane surface. As a general rule, membranes with higher hydrophilic and smoother surface usually have higher antifouling property. Furthermore, in some cases, static BSA adsorption test was also conducted to figure out the amount of protein adsorbed onto membrane surface after modification and to verify the improvement of antifouling property (Yi et al. 2010b). Static protein adsorption could also be used for evaluating membrane blood compatibility.

5.15 Anticoagulation Properties of Membranes

After modify PES membrane to improve its surface, characterisation of anticoagulation and other biological properties is necessary because PES has been widely used in various application, particularly biomedical field and wastewater treatment.

Plasma recalcification time (PRT), thromboplastic time (APTT), activated partial prothrombin time (PT) and the whole blood clotting time (WBCT) are frequently used as anticoagulation indices (Li et al. 2010, 2012; Zhao et al. 2011b; Huang et al. 2011). Additionally, albumin, cytocompatibility, platelet adhesion and fibrin adsorption are also biological indices used as biomaterials for modified PES membranes.

5.16 Membrane Stability

Membrane stability is a very important aspect after membrane modification. Long lifetime of the treated membrane during usage or application is the concern of many researchers to verify whether the modification process is applicable. Surprisingly, few research studies have been conducted on the stability of membrane after surface modification.

5.17 Stability After Physical Modification of PES Membrane

Yi et al. (2011) examined the stability of the modified PES membranes by washing them with water at temperature 60 °C for period 75 days after modification of polyethersulfone membranes with poly (propylene glycol) (PPG) (1000, 4000 and 8000 KDa). The stability of poly (propylene glycol) in membranes is confirmed by contact angle measurements. Two reasons have been suggested for the stability of PPG group into polymeric PES membrane: First, due to the characteristic of PPG (e.g. homopolymer, insoluble in water, Mw = 740 g/gmol), PPG is remained in membranes when membranes contact with water. The second reason is the good interaction (compatibility) between PES and PPG due to the bulky hydrophobic groups of PPG, which is assumed to be responsible for this interaction when the amphiphilic polymer PPG is added to the dope solution. Furthermore, the solubility parameters of PPG and PES are approximate (19.9 $(MPa)^{0.5}$ and 20.7 $(MPa)^{0.5}$ respectively) and this approximate solubility parameter will lead to effective interaction between PPG and PES when they are blended into membranes (Sung and Lin 1997; Niemelä et al. 1996; Yi et al. 2011).

Susanto and Ulbricht (2009) scrutinised the stability of the hydrophilic additives (PEG, PVP, pluronic) in the polymeric PES membrane network by washing the modified membranes physically in water at temperatures 20 and 40 °C and chemically in sodium hypochlorite solution (400 mg/L) for 10 days. As shown in Fig. 5.5, contact angle measurement showed that there is a significant increase in water contact angle of PES when PEG is used, indicating that PEG has thoroughly low stability in membranes. The reason for that is PEG is homopolymer, water-soluble polymer and can be leached out from the hydrophobic PES

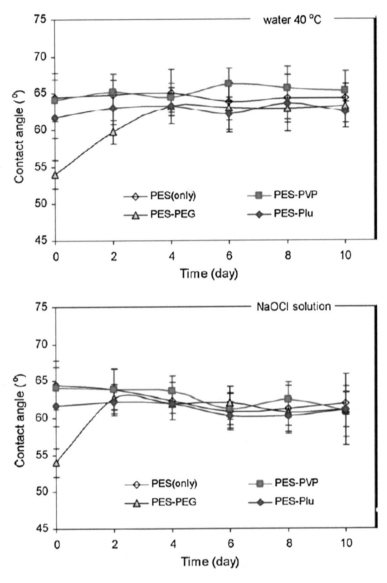

Fig. 5.5 Stability investigated by measuring the contact angle as a function of time (Susanto and Ulbricht 2009)

membranes during phase inversion process. (Susanto and Ulbricht 2009) also observed that IR absorbance for all the additives is slightly decreased (PEG, PVP, pluronic) as presented in Fig. 5.6. As it can be seen none of the additives above was entirely stable in the PES membrane. However, irrespective this loss, significant surface modification could still be seen as noted by either the existence of new peak

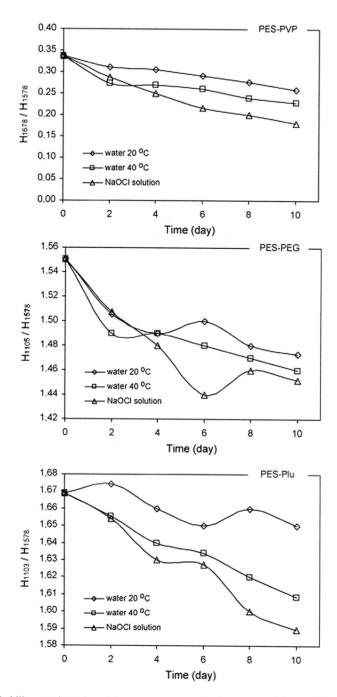

Fig. 5.6 Stability test investigated by measuring the IR absorbance of hydrophilic functional additive as a function of incubating time (Susanto and Ulbricht 2009)

for PES–PVP or increase in transmittance for PES–PEG and PES–Pluronic even
after incubating for 10 days in all studied potential cleaning agents.

5.18   Stability After Chemical Modification of PES
Membrane

Wei et al. (2012) modified PES membranes using CF_4 plasma process. Then, the
stability of the surface-treated PES hollow fibre membrane was scrutinised. A direct
contact membrane distillation experiment was conducted using hot salt solution
(NaCl) as feed at temperature 60.5 °C. The membrane module was tested in total
54 h for 9 days, 6 h per day. During the test period, the membrane module was in
contact with both feed and distillate. As can be seen in Fig. 5.7 the evaluation of
long life stability membrane showed that the membrane permeation (water flux) is
approximately stable and salt rejection was nearly 100 %.

Schulze et al. (2012) modified different types of polymer PES, PVDF, polysulfone and polyacrylonitrile membranes with aqueous solution of small molecules
(having hydrophilic groups, e.g. sulfonic acid, carboxylic acid, phosphonic acid,
alcohols, amine and zwitterionic components) and treated by low-energy electron
beam ion approach. The high stability of the modification showed permanent
membrane functionalities that are chemically attached.

Susanto and Ulbricht (2007) prepared polymeric PES membranes with high
antifouling-resistant. Two hydrophilic monomers including poly (ethylene glycol)
methacrylate   (PEGMA)   and   N,N-dimethyl-N-(2-methacryloyloxyethyl-N-
(3-sulfopropyl) ammonium betaine (SPE), with and without using cross-linker
monomer (N,N'-methylene bisacrylamide) (MBAA) have been grafted onto PES
membranes using photografting method. The stability of the grafted monomer was
evaluated chemically in sodium hypochlorite solution (NaOCl) (chemical agent).

Fig. 5.7 Stability
performance of the
surface-modified PES hollow
fiber membrane. Hot NaCl
solution: 4 wt% NaCl,
temperature 60.5 ± 0.2 °C,
2.0 m/s. Cold distillate water:
20 ± 0.5 °C, 0.68 m/s
(Wei et al. 2012)

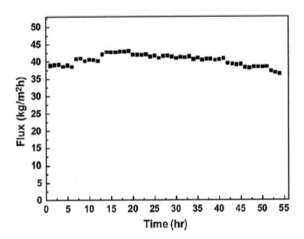

Fig. 5.8 Normalized
absorbance intensity (by
subtracting the absorbance
of the base membrane) at
wave numbers of 1725 cm^{-1}
for PEGMA- and 1727 cm^{-1}
for SPE-modified PES
membranes after soaking in
sodium hypochlorite solution
(Susanto and Ulbricht 2007)

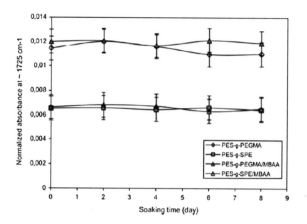

As shown in Fig. 5.8, ATR-IR absorbance for the introduced ester groups (relative
to groups in the PES backbone) remained constant after incubating for certain
period. (Susanto and Ulbricht 2007) also stated that such results are also verified by
measurements of contact angle and by elemental analyses (XPS). Furthermore,
(Susanto and Ulbricht 2007) argued that although numerous research studies had
stated previously that NaOCl could degrade the polymeric polysulfone or
polyethersulfone polymeric membranes, the results of their study research obviously indicated that either no loss or hydrolysis of the grafted monomer layers
happened within 8 days, verifying that the active layer of the photografted composite membranes is chemically stable.

References

Abu Seman MN, Khayet M, Bin Ali ZI, Hilal N (2010) Reduction of nanofiltration membrane
    fouling by UV-initiated graft polymerization technique. J Membr Sci 355(1–2):133–141
Bolong N, Ismail AF, Salim MR, Rana D, Matsuura T (2009) Development and characterization of
    novel charged surface modification macromolecule to polyethersulfone hollow fiber membrane
    with polyvinylpyrrolidone and water. J Membr Sci 331(1–2):40–49
Brayfield CA, Marra KG, Leonard JP, Tracy Cui X, Gerlach JC (2008) Excimer laser channel
    creation in polyethersulfone hollow fibers for compartmentalized in vitro neuronal cell culture
    scaffolds. Acta Biomater 4(2):244–255. doi:10.1016/j.actbio.2007.10.004
Cao X, Tang M, Liu F, Nie Y, Zhao C (2010) Immobilization of silver nanoparticles onto
    sulfonated polyethersulfone membranes as antibacterial materials. Colloids Surf, B 81(2):555–
    562. doi:10.1016/j.colsurfb.2010.07.057
Chu LY, Wang S, Chen WM (2005) Surface modification of ceramic-supported polyethersulfone
    membranes by interfacial polymerization for reduced membrane fouling. Macromol Chem
    Phys 206(19):1934–1940
Dejeu J, Lakard B, Fievet P, Lakard S (2009) Characterization of charge properties of an
    ultrafiltration membrane modified by surface grafting of poly(allylamine) hydrochloride.
    J Colloid Interface Sci 333(1):335–340. doi:10.1016/j.jcis.2008.12.069

Fontyn M, Bijsterbosch B, Van't Riet K (1986a) Chemical characterization of ultrafiltration membranes by spectroscopic techniques. J Membr Sci 36:141–145

Fontyn M, Bijsterbosch B, Van't Riet K (1986b) Chemical characterization of ultrafiltration membranes by spectroscopic techniques. J Membr Sci 36:141–145

Gaina C, Gaina V, Ionita D (2011) Chemical modification of chloromethylated polysulfones via click reactions. Polym Int 60(2):296–303. doi:10.1002/pi.2948

Hester JF, Banerjee P, Mayes AM (1999) Preparation of protein-resistant surfaces on poly (vinylidene fluoride) membranes via surface segregation. Macromolecules 32(5):1643–1650

Huang J, Xue J, Xiang K, Zhang X, Cheng C, Sun S, Zhao C (2011) Surface modification of polyethersulfone membranes by blending triblock copolymers of methoxyl poly(ethylene glycol)-polyurethane-methoxyl poly(ethylene glycol). Colloids Surf, B 88(1):315–324. doi:10. 1016/j.colsurfb.2011.07.008

Kochan J, Wintgens T, Wong JE, Melin T (2010) Properties of polyethersulfone ultrafiltration membranes modified by polyelectrolytes. Desalination 250(3):1008–1010. doi:10.1016/j.desal. 2009.09.092

Li Y, Chung TS, Kulprathipanja S (2005) Enhanced gas separation performance in polyether-sulfone (Pes)-modified zeolite mixed matrix membranes. In: Proceedings of AIChE annual meeting conference, p 2062

Li L, Yin Z, Li F, Xiang T, Chen Y, Zhao C (2010) Preparation and characterization of poly (acrylonitrile-acrylic acid-N-vinyl pyrrolidinone) terpolymer blended polyethersulfone membranes. J Membr Sci 349(1–2):56–64

Li L, Xiang T, Su B, Li H, Qian B, Zhao C (2012) Effect of membrane pore size on the pH-sensitivity of polyethersulfone hollow fiber ultrafiltration membrane. J Appl Polym Sci 123 (4):2320–2329. doi:10.1002/app.34902

Lin X, et al (2015) Composite ultrafiltration membranes from polymer and its quaternary phosphonium-functionalized derivative with enhanced water flux. J Membr Sci 482:67–75

Liu SX, Kim JT (2011) Characterization of surface modification of polyethersulfone membrane. J Adhes Sci Technol 25(1–3):193–212. doi:10.1163/016942410X503311

Liu SX, Kim JT, Kim S, Singh M (2009a) The effect of polymer surface modification via interfacial polymerization on polymer-protein interaction. J Appl Polym Sci 112(3):1704–1715. doi:10.1002/app.29606

Liu Z, Deng X, Wang M, Chen J, Zhang A, Gu Z, Zhao C (2009b) BSA-modified polyethersulfone membrane: preparation, characterization and biocompatibility. J Biomater Sci Polym Ed 20(3):377–397. doi:10.1163/156856209X412227

Low Z-X, Razmjou A, Wang K, Gray S, Duke M, Wang H (2014) Effect of addition of two-dimensional ZIF-L nanoflakes on the properties of polyethersulfone ultrafiltration membrane. J Membr Sci 460:9–17. doi:10.1016/j.memsci.2014.02.026

Mansourpanah Y, Madaeni SS, Adeli M, Rahimpour A, Farhadian A (2009a) Surface modification and preparation of nanofiltration membrane from polyethersulfone/polyimide blend-use of a new material (polyethyleneglycol-triazine). J Appl Polym Sci 112(5):2888–2895. doi:10.1002/ app.29821

Mansourpanah Y, Madaeni SS, Rahimpour A, Farhadian A, Taheri AH (2009b) Formation of appropriate sites on nanofiltration membrane surface for binding TiO$_2$ photo-catalyst: performance, characterization and fouling-resistant capability. J Membr Sci 330(1–2):297–306

Mansourpanah Y, Madaeni SS, Rahimpour A, Kheirollahi Z, Adeli M (2010) Changing the performance and morphology of polyethersulfone/polyimide blend nanofiltration membranes using trimethylamine. Desalination 256(1–3):101–107. doi:10.1016/j.desal.2010.02.006

Millesime L, Amiel C, Chaufer B (1994) Ultrafiltration of lysozyme and bovine serum albumin with polysulfone membranes modified with quaternized polyvinylimidazole. J Membr Sci J Membr Sci 89:223–234

Mosqueda-Jimenez DB, Narbaitz RM, Matsuura T (2004) Manufacturing conditions of surface-modified membranes: effects on ultrafiltration performance. Sep Purif Technol 37 (1):51–67. doi:10.1016/j.seppur.2003.07.003

Mosqueda-Jimenez DB, Narbaitz RM, Matsuura T (2006) Effects of preparation conditions on the surface modification and performance of polyethersulfone ultrafiltration membranes. J Appl Polym Sci 99(6):2978–2988. doi:10.1002/app.22993

Mu LJ, Zhao WZ (2009) Hydrophilic modification of polyethersulfone porous membranes via a thermal-induced surface crosslinking approach. Appl Surf Sci 255(16):7273–7278. doi:10.1016/j.apsusc.2009.03.081

Niemelä S, Leppänen J, Sundholm F (1996) Structural effects on free volume distribution in glassy polysulfones: molecular modelling of gas permeability. Polymer 37(18):4155–4165. doi:10.1016/0032-3861(96)00241-8

Pazokian H, Jelvani S, Mollabashi M, Barzin J, Azizabadi Farahani G (2011) ArF laser surface modification of polyethersulfone film: Effect of laser fluence in improving surface biocompatibility. Appl Surf Sci 257(14):6186–6190. doi:10.1016/j.apsusc.2011.02.028

Peeva PD, Million N, Ulbricht M (2012) Factors affecting the sieving behavior of anti-fouling thin-layer cross-linked hydrogel polyethersulfone composite ultrafiltration membranes. J Membr Sci 390–391:99–112

Pieracci J, Crivello JV, Belfort G (1999) Photochemical modification of 10 kDa polyethersulfone ultrafiltration membranes for reduction of biofouling. J Membr Sci 156(2):223–240

Rahimpour A (2011) UV photo-grafting of hydrophilic monomers onto the surface of nano-porous PES membranes for improving surface properties. Desalination 265(1–3):93–101

Rahimpour A, Madaeni SS, Taheri AH, Mansourpanah Y (2008) Coupling TiO_2 nanoparticles with UV irradiation for modification of polyethersulfone ultrafiltration membranes. J Membr Sci 313(1–2):158–169

Rana D, Matsuura T (2010) Surface modifications for antifouling membranes, Chem. Rev. 110:2448–2471

Razmjou A, Mansouri J, Chen V (2011) The effects of mechanical and chemical modification of TiO_2 nanoparticles on the surface chemistry, structure and fouling performance of PES ultrafiltration membranes. J Membr Sci 378(1–2):73–84

Saha NK, Balakrishnan M, Ulbricht M (2009) Fouling control in sugarcane juice ultrafiltration with surface modified polysulfone and polyethersulfone membranes. Desalination 249 (3):1124–1131. doi:10.1016/j.desal.2009.05.013

Schulze A, Marquardt B, Went M, Prager A, Buchmeiser MR (2012) Electron beam-based functionalization of polymer membranes. Water Sci Technol 65(3):574–580. doi:10.2166/wst.2012.890

Sung PH, Lin CY (1997) Polysiloxane modified epoxy polymer network—II. Dynamic mechanical behavior of multicomponent graft-IPNs (epoxy/polysiloxane/polypropylene glycol). Eur Polymer J 33(3):231–233

Susanto H, Ulbricht M (2007) Photografted thin polymer hydrogel layers on PES ultrafiltration membranes: characterization, stability, and influence on separation performance. Langmuir 23 (14):7818–7830

Susanto H, Ulbricht M (2009) Characteristics, performance and stability of polyethersulfone ultrafiltration membranes prepared by phase separation method using different macromolecular additives. J Membr Sci 327(1–2):125–135. doi:10.1016/j.memsci.2008.11.025

Tur E, Onal-Ulusoy B, Akdogan E, Mutlu M (2012) Surface modification of polyethersulfone membrane to improve its hydrophobic characteristics for waste frying oil filtration: radio frequency plasma treatment. J Appl Polym Sci 123(6):3402–3411. doi:10.1002/app.34400

Van der Bruggen B (2009) Comparison of redox initiated graft polymerisation and sulfonation for hydrophilisation of polyethersulfone nanofiltration membranes. Eur Polymer J 45(7):1873–1882. doi:10.1016/j.eurpolymj.2009.04.017

Vrijenhoek EM, Hong S, Elimelech M (2001) Influence of membrane surface properties on initial rate of colloidal fouling of reverse osmosis and nanofiltration membranes. J Membr Sci 188 (1):115–128. doi:10.1016/S0376-7388(01)00376-3

Wei X, Zhao B, Li XM, Wang Z, He BQ, He T, Jiang B (2012) CF 4 plasma surface modification of asymmetric hydrophilic polyethersulfone membranes for direct contact membrane distillation. J Membr Sci 407–408:164–175. doi:10.1016/j.memsci.2012.03.031

Wu TM, Yang SH (2006) Surface characterization and barrier properties of plasma-modified polyethersulfone/layered silicate nanocomposites. J Polym Sci, Part B: Polym Phys 44 (22):3185–3194. doi:10.1002/polb.20961

Yi Z, Zhu LP, Xu YY, Li XL, Yu JZ, Zhu BK (2010a) F127-based multi-block copolymer additives with poly(N, N-dimethylamino-2-ethyl methacrylate) end chains: the hydrophilicity and stimuli-responsive behavior investigation in polyethersulfone membranes modification. J Membr Sci 364(1–2):34–42. doi:10.1016/j.memsci.2010.07.045

Yi Z, Zhu LP, Xu YY, Zhao YF, Ma XT, Zhu BK (2010b) Polysulfone-based amphiphilic polymer for hydrophilicity and fouling-resistant modification of polyethersulfone membranes. J Membr Sci 365(1–2):25–33. doi:10.1016/j.memsci.2010.08.001

Yi Z, Zhu L, Xu Y, Jiang J, Zhu B (2011) Polypropylene glycol: the hydrophilic phenomena in the modification of polyethersulfone membranes. Ind Eng Chem Res 50(19):11297–11305. doi:10. 1021/ie201238c

Zhang J, Zhang M, Zhang K (2014) Fabrication of poly(ether sulfone)/poly(zinc acrylate) ultrafiltration membrane with anti-biofouling properties. J Membr Sci 460:18–24

Zhao C, Xue J, Ran F, Sun S (2013a) Modification of polyethersulfone membranes—a review of methods. Prog Mater Sci 58(1):76–150

Zhao W, He C, Wang H, Su B, Sun S, Zhao C (2011a) Improved antifouling property of polyethersulfone hollow fiber membranes using additive of poly(ethylene glycol) methyl ether-b-poly(styrene) copolymers. Ind Eng Chem Res 50(6):3295–3303. doi:10.1021/ ie102251v

Zhao W, Huang J, Fang B, Nie S, Yi N, Su B, Li H, Zhao C (2011b) Modification of polyethersulfone membrane by blending semi-interpenetrating network polymeric nanoparticles. J Membr Sci 369(1–2):258–266

Zhao, Yong-Hong, Qian, Yan-Ling, Zhu, Bao-Ku, et al (2008c) 'Modification of porous poly (vinylidene fluoride) membrane using amphiphilic polymers with different structures in phase inversion process', J Membr Sci 310(1–2):567–76

Printed in the United States
By Bookmasters